T0334375

HEAT EXCHANGER DESIGN GUIDE

HEAT EXCHANGER DESIGN GUIDE

A Practical Guide for Planning, Selecting and Designing of Shell and Tube Exchangers

M. NITSCHE AND R.O. GBADAMOSI

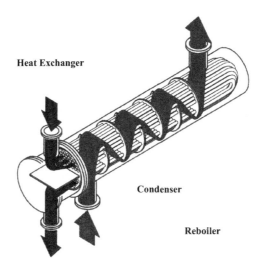

Heat Exchanger

Condenser

Reboiler

With numerous practical Examples

Amsterdam • Boston • Heidelberg • London • New York • Oxford
Paris • San Diego • San Francisco • Singapore • Sydney • Tokyo
Butterworth Heinemann is an imprint of Elsevier

Butterworth Heinemann is an imprint of Elsevier
The Boulevard, Langford Lane, Kidlington, Oxford OX5 1GB, UK
225 Wyman Street, Waltham, MA 02451, USA

Copyright © 2016 Elsevier Inc. All rights reserved.

No part of this publication may be reproduced or transmitted in any form or by any means, electronic or
mechanical, including photocopying, recording, or any information storage and retrieval system, without
permission in writing from the publisher. Details on how to seek permission, further information about
the Publisher's permissions policies and our arrangements with organizations such as the Copyright
Clearance Center and the Copyright Licensing Agency, can be found at our website: www.elsevier.com/
permissions.

This book and the individual contributions contained in it are protected under copyright by the Publisher
(other than as may be noted herein).

Notices
Knowledge and best practice in this field are constantly changing. As new research and experience
broaden our understanding, changes in research methods, professional practices, or medical treatment
may become necessary.

Practitioners and researchers must always rely on their own experience and knowledge in evaluating and
using any information, methods, compounds, or experiments described herein. In using such information
or methods they should be mindful of their own safety and the safety of others, including parties for
whom they have a professional responsibility.

To the fullest extent of the law, neither the Publisher nor the authors, contributors, or editors, assume any
liability for any injury and/or damage to persons or property as a matter of products liability, negligence or
otherwise, or from any use or operation of any methods, products, instructions, or ideas contained in the
material herein.

ISBN: 978-0-12-803764-5

British Library Cataloguing-in-Publication Data
A catalogue record for this book is available from the British Library

Library of Congress Cataloging-in-Publication Data
A catalog record for this book is available from the Library of Congress

For information on all Butterworth Heinemann publications
visit our website at http://store.elsevier.com/

Working together
to grow libraries in
developing countries

www.elsevier.com • www.bookaid.org

Publisher: Joe Hayton
Acquisition Editor: Fiona Geraghty
Editorial Project Manager: Cari Owen
Production Project Manager: Susan Li
Designer: Victoria Pearson

Typeset by TNQ Books and Journals
www.tnq.co.in

Printed and bound in the United States of America

CONTENTS

FOREWORD

Dear Reader,

This book is not an academic treatise but rather a book for solving daily practical problems easily and for the illustration of essential influencing variables in the design of heat exchangers, condensers, or evaporators.

All calculations are explained with examples that I am using in my seminars since several years.

In this book, you will be shown how to proceed in the design of a heat exchanger in the daily practice, how to determine the effective temperature difference for the heat transfer, and how to calculate the heat transfer coefficient using simple equations.

The most important influence parameters for the heat transfer coefficient are introduced. Different calculation models are compared. It is shown how to calculate the required dew point and bubble point lines for mixtures.

From the wide range of published calculation methods, I have chosen the models that can simply be calculated using the hand calculator and deliver sure results. I refer to the models which I have chosen from several existing literatures as Nitsche methods because I recommend these.

During the time from 1966 to 2007, I designed, planned, and built several chemical plants: distillation plants with evaporators, condensers, and heat exchangers, for fatty alcohols, fatty acids, nitrochlorobenzenes, amine and hydrocarbons and tar oils; storage tanks and vessels with filling stations for tank truck, rail tank car, and barrels; stirred tank plants for reactions with decanters, centrifuges, and filters; plants for exhaust air purification and gasoline recovery, methanol and ethyl acetate etc.; stripper for of sour water purification or for methyl isobutyl ketone recovery.

Since 1980, I report in seminars about piping, heat exchangers, and chemical plants about my practical experiences in the Design and Planning. At the latest, I realized during plant startup what I wrongly calculated or what I wrongly planed. This book should help you to minimize the mistakes in the design of heat exchangers.

Hamburg
Dr Manfred Nitsche

CHAPTER 1

Heat Exchanger Design

Contents

In heat exchanger design the required heat exchanger area A (m^2) is determined for a certain heat load Q (W) at a given temperature gradient Δt (°C).

$$A = \frac{Q}{U \times \Delta t} \left(\text{m}^2\right)$$

The overall heat transfer coefficient, U, is calculated as follows:

$$\frac{1}{U} = \frac{1}{\alpha_i} + \frac{1}{\alpha_o} + \frac{s}{\lambda} + f_i + f_o$$

f_i = inner fouling factor (m^2 K/W)
f_o = outer fouling factor (m^2 K/W)
U = overall heat transfer coefficient (W/m^2 K)
s = tube wall thickness (m)
λ = thermal conductivity of the tube material (W/m K)
α_i = inner heat transfer coefficient in the tubes (W/m^2 K)
α_o = outer heat transfer coefficient on the shell side (W/m^2 K)

Figure 1.1 shows which overall heat resistances have to be overcome and how the temperature profile in a heat exchanger looks like.

The overall heat transfer coefficients and the temperature profile will be calculated in Chapter 6.

Reference values for heat transfer coefficients and overall heat transfer coefficients are listed in Table 1.1.

© 2016 Elsevier Inc.
All rights reserved.

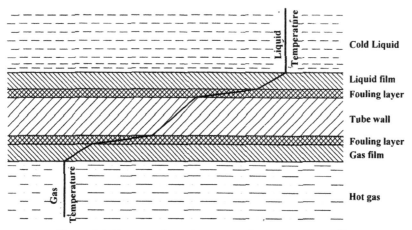

Figure 1.1 Heat transfer resistances and temperature profile.

1.1 PROCEDURE IN HEAT EXCHANGER DESIGN

In order to calculate the convective heat transfer coefficients, the Reynolds number is needed.

The heat transfer coefficients, α, are dependent on the Reynolds number, Re, hence the flow velocity, w, on the tube and shell side, respectively.

$$\textit{Tube side}: \alpha \propto w^{0.8} \qquad \textit{Shell side}: \alpha \propto w^{0.6}$$

Therefore the cross-sectional areas must be known in order to determine the flow velocities and the Reynolds numbers. For an existing heat exchanger, this is not a problem if a drawing is available. In the case of the design of a new heat exchanger, the flow cross sections are not known. So, initially an estimation of the required area has to be done and then an appropriate equipment has to be selected.

For the selection, the following criteria should be applied:
- The flow velocity on both sides should be in the order of 0.5—1 m/s for liquids and in the range of 15—20 m/s for gases.
- The required heat exchanger area should be achieved with tube length of 3—6 m.

In Figure 1.3 the flow chart for the heat exchanger design is provided [1].

In the following the procedure of heat exchanger design is explained in some more detail:

1. Determine flow rates, temperatures, and the fluid property data
2. Determination of the heat loads on tube and shell side

$$\textit{Shellside}: \quad Q_{S\,req} = M_S \times c_S \times (T_1 - T_2)(W)$$
$$\textit{Tubeside}: \quad Q_{T\,req} = M_T \times c_T \times (t_2 - t_1)(W)$$

Table 1.1 Reference values for heat transfer coefficients, α

Natural convection		α (W/m^2 K)
Gases at atmospheric pressure		4—6
Oil (viscosity = 100 mm^2/s)		10—20
Water		250—500
Hydrocarbons, low viscosity		170—300

Condensation		
Steam		5000—10000
Organic solvents		1000—3000
Light oils		1000—1500
Heavy oils (vacuum)		100—300

Vaporization		
Water		4000—10000
Organic solvents		1000—2500
Light oils		700—1400

Flowing media		
Atmospheric gases		40—200
Gases under pressure		150—300
Organic solvents		300—1000
Water		2500—4000

Guiding values for overall heat transfer coefficients, U

Condensation		U (W/m^2 K)
Water	Water	1000—2000
	Organic solvent	600—1000
	Organic solvent + inert gases	100—500
	Heavy hydrocarbons	50—200

Evaporation		
Steam	Water	2000—4000
	Organic solvent	500—1000
	Light oils	250—800
	Heavy oils	120—400

Flowing media		
Steam	Water	1500—4000
	Organic solvents	600—1000
	Gases	30—250
Water	Water	1000—2000
	Organic solvents	250—800
	Gases	15—300
Organic solvents	Organic solvents	100—300

Table 1.2 Geometric data of heat exchangers according to DIN 28,184, part 1, for 25 × 2 tubes with 32 mm triangular pitch

Typ nr.	DN	Z	Da (mm)	B (mm)	n	AE (mm²)	AR (mm²)	AS (m²/m)	f_w	VR (m³/h)	VM (m³/h)
1	150	2	168	30	14	1770	2425	1.1	0.251	8.73	6.37
2	200	2	219	40	26	2288	4503	2	0.366	16.21	8.24
3	250	2	273	50	44	5520	7620	3.5	0.259	27.43	19.87
4	300	2	324	60	66	5088	11,430	5.2	0.375	41.15	18.32
5	350	2	355	70	76	6230	13,162	6	0.397	47.38	22.43
6	350	4	355	70	68	6230	5888	5.3	0.381	21.20	22.43
7	400	2	406	80	106	11,072	18,357	8.3	0.319	66.09	39.86
8	400	4	406	80	88	9072	7620	6.9	0.382	27.43	32.66
9	500	2	508	100	180	14,600	31,172	14.1	0.360	112.22	52.56
10	500	4	508	100	164	12,100	14,201	12.9	0.430	51.12	43.56
11	600	2	600	120	258	19,560	44,681	20.3	0.389	160.85	70.42
12	600	8	600	120	232	22,560	10,044	18.2	0.348	36.16	81.22
13	700	2	700	140	364	22,260	63,038	28.6	0.456	226.94	80.14
14	700	8	700	140	324	25,760	14,028	25.4	0.395	50.50	92.74
15	800	2	800	160	484	29,440	83,819	38	0.454	301.75	105.98
16	800	8	800	160	432	37,440	18,703	33.9	0.367	67.33	134.78
17	900	2	900	180	622	41,400	107,718	48.9	0.407	387.79	149.04
18	900	8	900	180	556	41,400	24,072	43.7	0.416	86.66	149.04
19	1000	2	1000	200	776	46,000	134,388	61	0.452	483.80	165.60
20	1000	8	1000	200	712	56,000	30,826	55.9	0.373	110.97	201.60
21	1100	2	1100	220	934	55,220	161,750	73.4	0.460	582.30	198.79
22	1100	8	1100	220	860	60,720	37,234	67.5	0.420	134.05	218.59
23	1200	2	1200	240	1124	72,240	194,655	88.3	0.416	700.76	260.06
24	1200	8	1200	240	1048	78,240	45,373	82.3	0.390	163.34	281.66

DN = Nominal shell diameter; Z = number of tube passes; Da = shell diameter (mm); B = baffle spacing (mm); n = number of tubes; AE = shell-side flow cross section (mm²); AR = tube-side flow cross section (mm²); AS = Heat exchanger area per m tube length (m²/m); VR = Required flow rate for 1 m/s flow velocity on the tube side (m³/h); VM = Required flow rate for 1 m/s flow velocity at the shell side (m³/h).

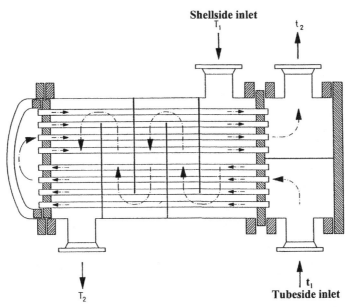

Figure 1.2 Heat exchanger for convective heat transfer.

For condensers and evaporators, the condensation and the vaporization enthalpies need to be considered.

3. Calculation of the corrected effective temperature difference (CMTD) for the heat load

First, the logarithmic mean temperature difference (LMTD) is determined for ideal countercurrent flow.

Most heat exchangers have multiple passes in order to increase flow velocity in the tubes.

For example, the heat exchanger in Figure 1.2 has two passes. The medium on the tube side flows forward and backward. In one tube pass the fluid flows cocurrent to the flow on the shell side. In the other pass the flow is in countercurrent to the shell-side flow.

Ideal countercurrent flow does not occur. The driving temperature gradient is worse. Therefore for heat exchangers with multiple passes, due to the nonideal countercurrent flow, the temperature efficiency factor, F, must be calculated. F *should be* >0.75!

Using the temperature efficiency factor, F, the CMTD is determined.

$$\text{CMTD} = F \times \text{LMTD}.$$

The calculation of the effective temperature difference is shown in Chapter 2.

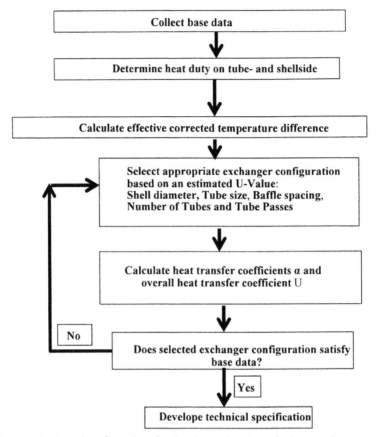

Figure 1.3 Procedure flow chart for the thermal design of a heat exchanger [1].

With nonlinear condensation or evaporation curves, the average weighted temperature difference must be determined.

Zones with approximate linear range of the condensation temperature are established for which the CMTDs are determined. Finally, the weighted average of the effective temperature differences in the zones is determined.

4. Estimation of the required heat exchanger area

For the calculated heat load and the available effective temperature difference, the required heat exchanger area is estimated using the estimated overall heat transfer coefficient, U, from Table 1.1:

$$A = \frac{Q_{req}}{U \times CMTD} \ (m^2)$$

5. Selection of an appropriate equipment from Table 1.2 for the required heat exchange area, A, from column, AS, with the area of the heat exchanger per m length (m^2/m)

6. Determination of the flow velocity using the columns, VR and VM, in Table 1.2
   The volumetric flows on the tube and shell side are listed in columns, VR and VM, which are required for a flow velocity of 1 m/s.

Example 1: Selection of an appropriate heat exchanger for a required area $A = 55$ m^2

Tube-side flow rate $V_{tube} = 40$ m^3/h, shell-side flow rate $V_{shell} = 80$ m^3/h

Selected: Type 12 with 18.2 m^2/m tube length and 232 tubes in 8 passes, DN 600
Tube length $= 4$ m, Heat exchanger area $= 4 \times 18.2 = 72.8$ m^2 (32% excess)
VR $= 36.16$ m^3/h for 1 m/s and VM $= 81.22$ m^3/h for 1 m/s from Table 1.2
Determination of the flow velocity, w_t, on the tube side:

$$w_t = \frac{V_{tube}}{VR} \times 1 = \frac{40}{36.16} \times 1 = 1.1 \text{ m/s}$$

Determination of the flow velocity, w_{sh}, on the shell side:

$$w_{sh} = \frac{V_{shell}}{VM} \times 1 = \frac{80}{81.22} \times 1 = 0.985 \text{ m/s}$$

Primary condition for a good convective heat transfer is an adequately high flow velocity.
That is why both columns for VR and VM in Table 1.2 are important.
Figures 1.4 and 1.5 show that the heat transfer coefficients increase with rising flow velocities.
On the shell side, the baffle spacing, B, can be shortened, for instance, if the flow velocity shall be increased in order to achieve a better heat transfer coefficient.
Since the flows on which the design is based are known the flow velocities can be easily determined with VR and VM (see Example 1).

---

7. Calculation of the convective heat transfer coefficients on the tube and shell side
   The Reynolds number can be calculated once the flow velocity is determined.
   With convective heat transfer, the heat transfer coefficient is dependent on the Reynolds number, Re, and the Prandtl number, Pr (see Chapter 3).

$$\alpha = \frac{\text{Const} \times \text{Re}^m \times \text{Pr}^{0.33} \times \lambda}{d} \qquad \text{Re} = \frac{w \times d}{\nu} \qquad \text{Pr} = \frac{\nu \times c \times \rho \times 3600}{\lambda}$$

$\nu =$ kinematic viscosity (mm^2/s)
$d =$ tube diameter (m)
$w =$ flow velocity (m/s)
$\lambda =$ heat conductivity (W/m K)
$\rho =$ density (kg/m^3)
$c =$ specific heat (Wh/kg K)

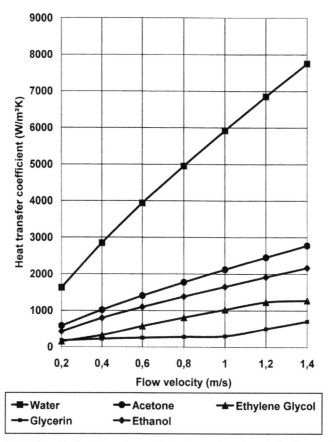

Figure 1.4 Tube-side heat transfer coefficient as a function of the flow velocity.

The Prandtl number, Pr, is only dependent on the physical properties, that is, density, viscosity, heat conductivity, and specific heat.

The calculated heat transfer coefficient, α_i, in the tubes is converted with respect to the outer area because the heat exchanger area refers to the tube outer diameter.

$$\alpha_{io} = \alpha_i \times \frac{d_i}{d_o} \left(\text{W}/\text{m}^2 \, \text{K} \right)$$

8. *Determination of the overall heat transfer coefficient, U*, considering the resistance of the tube wall and potential coating and fouling layers

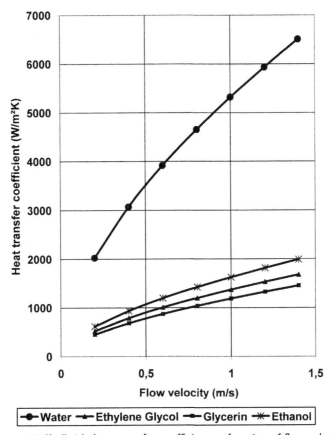

Figure 1.5 Shell-side heat transfer coefficient as function of flow velocity.

9. *Calculation of the actual heat load* of the selected heat exchanger and comparison with the U-value for the required heat load

$$Q = U \times A \times \text{CMTD} \ (W)$$

10. Checking the excess and the fouling reserve

$$\left(\frac{U}{U_{\text{req}}} \times 100\right) - 100 = \% \quad \text{reserve}$$

Example 2: Design of a water cooler

1. Required data

	Shell side	Tube side
Flow rate (kg/h)	50,000	36,800
Inlet temperature (°C)	65	15
Outlet temperature (°C)	50	35.4
Density ρ (kg/m^3)	983	994
Specific heat capacity, c (Wh/kg K)	1.16	1.16
Kinematic Viscosity, ν (mm^2/s)	0.47	0.92
Heat conductivity, λ (W/m K)	0.654	0.605
Fouling factor, f (m^2 K/W)	0.002	0.002

2. Determining the heat duty

Shell side $Q = m \times c \times \Delta t = 50{,}000 \times 1.16 \times (65.0 - 50) = 870{,}000$ W
Tube side $Q = m \times c \times \Delta t = 36{,}800 \times 1.16 \times (35.4 - 15) = 870{,}000$ W

3. Temperature difference, CMTD

The calculation follows the scheme described in Chapter 2. First the LMTD is determined for ideal countercurrent flow.

$$
\begin{array}{ccccc}
T_1 & \rightarrow & T_2 & 65.0 & \rightarrow & 50 \\
t_2 & \approx & t_1 & 35.4 & \approx & 15 \\
& & & 29.6 & & 35
\end{array}
$$

$$
\text{LMTD} = \frac{35 - 29.6}{\ln \frac{35}{29.6}} = 32.2 \ ^{\circ}\text{C}
$$

Since we intend to use a multipass heat exchanger, the temperature efficiency factor, F, for the correction of the logarithmic temperature gradients LMTD must be used.

F can be determined using Figure 1.6.

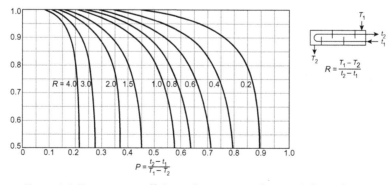

Figure 1.6 Temperature efficiency factor, F, as a function of P and R [3].

The parameters P and R are calculated from the temperatures T_1, T_2, t_1, and t_2.

$$P = 0.41 \quad R = 0.735$$

From the diagram in Figure 1.6, the value of F is read off at the intersection of P and R: $F = 0.948$

$$\text{CMTD} = F \times \text{LMTD} = 0.948 \times 32.2 = 30.55\,°C$$

4. Estimation of the required heat exchanger area, A

Heat load, $Q = 870{,}000$ W
Temperature gradient, CMTD $= 30.55\,°C$
Estimated overall heat transfer coefficient, $U = 716$ W/m^2 K
Required heat exchanger area, A, is given below:

$$A = \frac{Q}{U \times \text{CMTD}} = \frac{870{,}000}{716 \times 30.55} = 39.8\ \text{m}^2$$

5. Selection of an appropriate heat exchanger from Table 1.2

A heat exchanger type having the required area and sufficient flow velocities with the flows on the tube side and the shell side must be selected.

Chosen: type 10 DN 500 with 4 passes and 164 tubes of 25 × 2.

Determination of the heat exchanger area with 3 m tube length: $A = 3 \times 12.9 = 38.7$ m^2

6. Determination of the flow velocity using the columns VR and VM in Table 1.2

The volumetric flows on the tube and shell side are listed in columns, VR and VM, which are required for a flow velocity of 1 m/s.

Checking the tube side:

On the tube side there must be a flow of VR $= 51.12$ m^3/h for a flow velocity of 1 m/s.
In our example, the flow on the tube side is only $V_{tube} = 37$ m^3/h.
The resulting real flow velocity, w_t, is as follows:

$$w_t = \frac{V_{tube}}{VR} \times 1 = \frac{37}{51.12} \times 1 = 0.72\ \text{m/s}$$

Checking the shell side:

On the shell side VM $= 43.56$ m/h for 1 m/s with a baffle spacing, B $= 100$ mm.
For B $= 200$ mm the flow is $2 \times 43.56 = 87.1$ m^3/h for 1 m/s on the shell side.
In the example, the flow on the shell side, $V_{sh} = 50$ m/h.
This gives a real flow velocity, w_{sh}, as follows:

$$w_{sh} = \frac{V_{shell}}{VM} \times 1 = \frac{50}{87.1} \times 1 = 0.58\ \text{m/s}$$

7. Calculation of the heat transfer coefficients on the tube side and the shell side

Using the found flow velocities on tube and shell side, the Reynolds numbers can be determined.
The calculation of the heat transfer coefficients will be described in Chapter 3.

$$\textit{Tube-side:} \quad w_t = 0.72\ \text{m/s}\ \textit{für}\ 37\ \text{m}^3/\text{h}$$

$$\text{Re}\ w \times \frac{d}{\nu} = 0.72 \times \frac{0.021}{0.92 \times 10^{-6}} = 16{,}435 \quad \text{Pr} = \frac{c \times \nu \times \rho \times 3600}{\lambda} = 6.32$$

$$\text{Nu} = 0.023 \times 16{,}435^{0.8} \times 6.32^{0.33} = 99.67$$

$$\alpha_i = \frac{99.67 \times 0.605}{0.021} = 2872\ \text{W/m}^2\ \text{K} \quad \alpha_{io} = 2872 \times \frac{21}{25} = 2412\ \text{W/m}^2\ \text{K}$$

Shell-side: $w_{shell} = 0.58$ m/s for 50.85 m^3/h

$$Re = 0.58 \times \frac{0.025}{0.47 \times 10^{-6}} = 30,851 \quad Pr = \frac{0.47 \times 10^{-6} \times 1.16 \times 983 \times 3600}{0.654} = 2.95$$

$$Nu = 0.196 \times 30,851^{0.6} \times 2.95^{0.33} = 138.3$$

$$\alpha_o = \frac{138.3 \times 0.654}{0.025} = 3618 \, \text{W/m}^2 \, \text{K}$$

8. *Determination of the overall heat transfer coefficient, U,* considering the resistances of the tube wall and the fouling factor.

Tube wall thickness, $s = 0.002$ m; wall heat conductivity, $\lambda = 50$ W/m K; tube-side fouling, $f_i = 0.0002$; shell-side fouling, $f_o = 0.0002$

$$\frac{1}{U} = \frac{1}{\alpha_{io}} + \frac{1}{\alpha_o} + \frac{s}{\lambda} + f_i + f_o = \frac{1}{2412} + \frac{1}{3618} + \frac{0.002}{50} + 0.0004 = 0.00113$$

$$U = 884 \, \text{W/m}^2 \, \text{K}$$

9. Calculation of the actual heat load of the selected heat exchanger and comparison with the *U*-value for the required heat load, Q_{req}

$$Q = U \times A \times CMTD = 884 \times 38.7 \times 30.55 = 1,045,140 \, \text{W}$$

$$U_{req} = \frac{Q_{req}}{A \times CMTD} = \frac{870,000}{38.7 \times 30.55} = 736 \, \text{W/m}^2 \, \text{K}$$

10. Checking the excess area and the fouling reserve

$$Excess \, area = \left(\frac{U}{U_{req}} \times 100\right) - 100 = \left(\frac{884}{736} \times 100\right) - 100 = 20.1\%$$

The heat exchanger has an excess area of 20.1%.

$$f_{res} = \frac{U - U_{req}}{U \times U_{req}} = \frac{884 - 736}{884 \times 736} = 0.000227 \, \text{m}^2 \, \text{K/W}$$

The excess for the fouling is 0.000227 m^2 K/W.

1.2 INFORMATION ABOUT HEAT EXCHANGERS

1.2.1 Tube Pattern

A better heat transfer coefficient can be achieved through a triangular or rotated tube pattern in the tube bundle because the flow is directly against the tube. In a quadratic tube pattern, the medium flows between the tube rows and the heat transfer coefficient are worse than with a rotated tube pattern. If a severe fouling is expected on the shell side, for instance, in a kettle–type evaporator, the quadratic pattern must be chosen to facilitate the cleaning of the

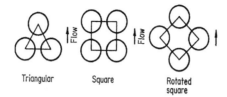

Figure 1.7 Tube arrangements in shell and tube heat exchangers (1.9).

shell side. A compromise is the rotated quadratic pitch. The heat transfer is better because the flow is directly against the tubes and cleaning can take place between the tube rows.

Normally, the triangular pitch is chosen because the heat transfer is good and with triangular pitch more tubes can fit in the shell. With triangular pitch the heat exchanger area is greater in a pipe shell (Figure 1.7).

1.2.2 Bypass and Leakage Streams

On the shell side the medium should flow through the tube bundle between the baffles.

If, however, large bypass lanes are installed, for instance, large gaps between the shell and the tube bundle, the product flows through the free lanes and this bypass flow does not take part in the heat exchange process.

If the holes in the baffles are much larger than the tube diameters, leakage streams through this gap results. The leakage stream flow does not take part in the heat exchange process.

In order to improve the efficiency of a heat exchanger, the bypass gap between the outer tube rows and the shell should be minimized or blocked with blind tubes in order to force the medium to flow across through the tube rows.

With floating head heat equipments which have a larger gap in the shell due to the conditional construction, sealing strips must block the outer gap (Figure 1.8).

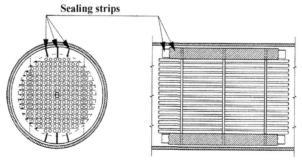

Figure 1.8 Sealing strips to minimize bypass stream.

Also free tube lanes through pass partitions in flow direction or with *U*-tube bundles must be avoided. The tolerances between the tubes and the baffle holes and between baffles and the shell should be as small as possible in order to minimize the bypass flow.

Recommended Tolerances: TEMA or DIN 28008, DIN 28182 and DIN 28185

1.2.3 Baffles

On the shell side, the flow cross section is given through the baffle spacing.

Some design forms are shown in Figure 3.12. Normally, segmental baffles are used (Figure 1.9).

For convective heat transfer on the shell side, the following design is recommended:

$$\text{Baffle spacing, } B = 0.2 \times D_i \qquad \text{Segmental height, } H = 0.2 \times D_i$$

With convective heat transfer an up—down flow at the shell side with horizontal baffle edges should be recommended because in this way stratified flow will be avoided.

For ventilation and draining, small slots can be arranged in the baffles which of course increase the bypass flows.

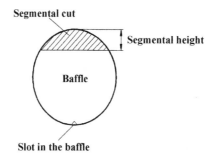

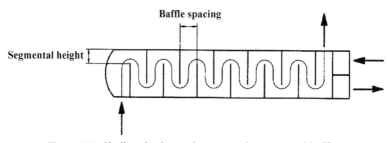

Figure 1.9 Shell and tube exchanger with segmental baffles.

In horizontal condensers with the vapors on the shell side, one should try to achieve a horizontal left–right flow with vertical baffle edges so that the condensate at the bottom can safely drain out at the bottom and the noncondensables can flow out at the top.

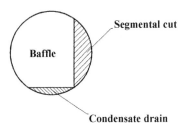

The baffle spacings can be adjusted for the decreasing vapor stream, that is, smaller with increasing flow path. This is important with vapors containing inert gases which must be convectively cooled before condensing.

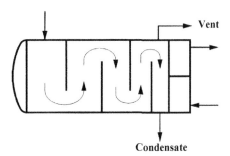

1.2.4 Technical Remarks

Which product should be to the tube or shell side?
    Through the tubes: the medium with higher pressure, corrosive media and products with high fouling tendency,
    On the shell side: viscous products.
How are the different expansions accommodated?
    Through an expansion joint in the shell of fixed tube sheet bundles,
    Through a *U*-tube construction with free tube expansion,
    With a floating head for the free expansion.
The different possibilities are shown in Figure 1.10.

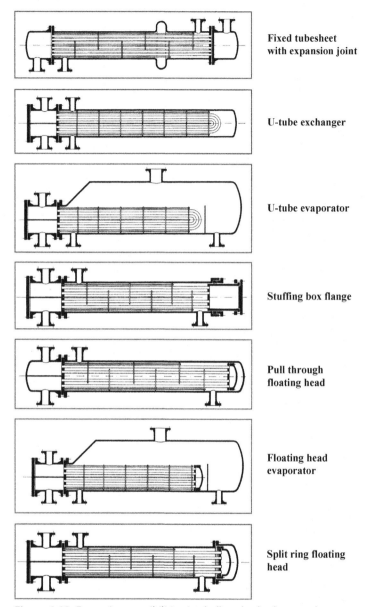

Fixed tubesheet
with expansion joint

U-tube exchanger

U-tube evaporator

Stuffing box flange

Pull through
floating head

Floating head
evaporator

Split ring floating
head

Figure 1.10 Expansion possibilities in shell and tube heat exchangers.

1.2.5 Selection of a Shell and Tube Exchanger

An overview of the different heat exchanger types according to TEMA [3] for different
bonnets, shells, and baffle chambers is given in Figure 1.11.

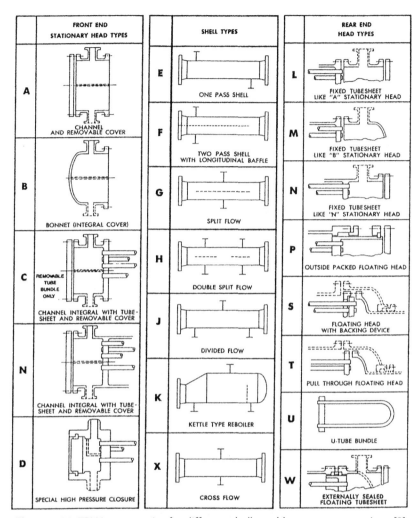

Figure 1.11 TEMA systematics for different shells and bonnets constructions [3].

With the help of the logic flowchart in Figure 1.12 the required heat exchanger type can be determined for the given problem definition. Thermal expansion stresses and shell-side fouling are of overriding importance.

1.2.5.1 Which Heat Exchanger Types Can Be Cleaned?

Tube-side cleaning: TEMA-types, AEL and NEN, with tube diameters >20 mm
    Shell-side cleaning:
    Removable bundle with square pitch: TEMA-types, CEU + CES + AES

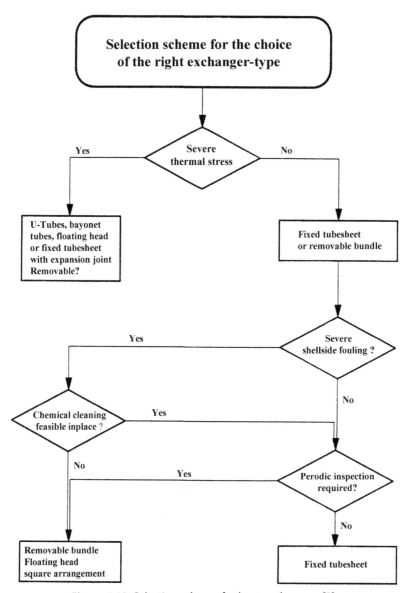

Figure 1.12 Selection scheme for heat exchangers [2].

NOMENCLATURE

U Calculated overall heat transfer coefficient (W/m^2 K)
U_{req} Required overall heat transfer coefficient (W/m^2 K)
M_S Shell-side flow rate (kg/h)
c_S Specific heat capacity shell side (Wh/kg K)

M_T Tube-side flow rate (kg/h)
c_T Specific heat capacity tube side (Wh/kg K)
Q_{req} Required heat load (W)
w_{sh} Shell-side flow velocity (m/s)
w_t Tube-side flow velocity (m/s)
Σf Fouling reserve (m^2 K/W)
α_{io} Heat transfer coefficient α_i in the tube relative to the outer area (W/m^2 K)
α_o Heat transfer coefficient of the shell-side (W/m^2 K)
s Tube wall thickness (m)
λ Thermal conductivity of the tube material (W/m K)
f_i Tube-side fouling factor inside the tube (m^2 K/W)
f_o Tube fouling factor at the outer side of the tube (m^2 K/W)
CMTD Corrected effective mean temperature difference $= F \times$ LMTD ($^\circ$C)
LMTD Logarithmic mean temperature difference for countercurrent ($^\circ$C)

REFERENCES

[1] A.E. Jones, Thermal design of the shell and tube, Chem. Eng. 109 (2002) 60–65.
[2] A. Devore, J. Vago, G.J. Picozzi, Specifying and selecting, Chem. Eng. 87 (1980) 133–148.
[3] TEMA, Standards of the Tubular Exchanger Manufacturers Association, eighth ed., 1999.

CHAPTER 2

Calculations of the Temperature Differences LMTD and CMTD

Contents

2.1 LOGARITHMIC MEAN TEMPERATURE DIFFERENCE FOR IDEAL COUNTERCURRENT FLOW

The logarithmic mean temperature difference (LMTD) for ideal countercurrent flow is determined from the two temperature differences Δt_1 and Δt_2.

$$\text{LMTD} = \frac{\Delta t_1 - \Delta t_2}{\ln \frac{\Delta t_1}{\Delta t_2}} \quad \Delta t_1 = T_1 - t_2 \quad \Delta t_2 = T_2 - t_1$$

T_1 = shell-side inlet temperature (°C)
T_2 = shell-side outlet temperature (°C)
t_1 = tube-side inlet temperature (°C)
t_2 = tube-side outlet temperature (°C)

Example 1: Calculation of the logarithmic mean temperature difference

Tube side: $t_1 = 30\,°C$ $t_2 = 60\,°C$
Shell side: isothermal heating   *Nonisothermal heating*

$T_1 = T_2 = 100\,°C$ $T_1 = 100\,°C$ $T_2 = 80\,°C$

$$\begin{array}{ccc} 100 & \rightarrow & 100 \\ \underline{60} & \approx & \underline{30} \\ \Delta t_1 = 40 & & \Delta t_2 = 70 \end{array} \qquad \begin{array}{ccc} 100 & \rightarrow & 80 \\ \underline{60} & \approx & \underline{30} \\ \Delta t_1 = 40 & & \Delta t_2 = 50 \end{array}$$

$\text{LMTD} = \frac{40-70}{\ln \frac{40}{70}} = 53.6\,°C$ $\text{LMTD} = \frac{40-50}{\ln \frac{40}{50}} = 44.8\,°C$

© 2016 Elsevier Inc.
All rights reserved.

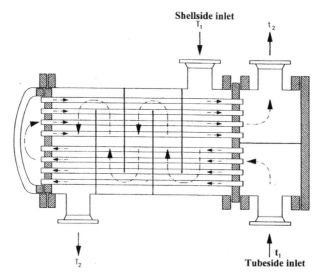

Figure 2.1 Heat exchanger TEMA—type E with two tube passes.

2.2 CORRECTED TEMPERATURE DIFFERENCE FOR MULTIPASS HEAT EXCHANGER

In Chapter 1, it was already pointed out that with a multipass heat exchanger, there is no ideal countercurrent but a mixture of co- and countercurrent flow.

This makes the effective temperature gradient worse.

In Figures 2.2 and 2.3, temperature efficiency diagrams are shown, it is clear that the driving temperature difference in the one-pass equipment with ideal countercurrent is better than in the two-pass heat exchanger.

Procedure for the determination of the corrected effective mean temperature difference (CMTD) for heat exchangers with nonideal countercurrent flow:

First, the LMTD is calculated from the inlet and outlet temperatures on the tube and shell side.

With nonideal countercurrent, the LMTD must be corrected with a temperature efficiency factor F.

$$CMTD = F \times LMTD$$

$F =$ temperature efficiency factor
LMTD $=$ logarithmic mean temperature difference ($^{\circ}$C) for countercurrent flow
CMTD $=$ corrected effective mean temperature difference ($^{\circ}$C) for multipass heat exchangers

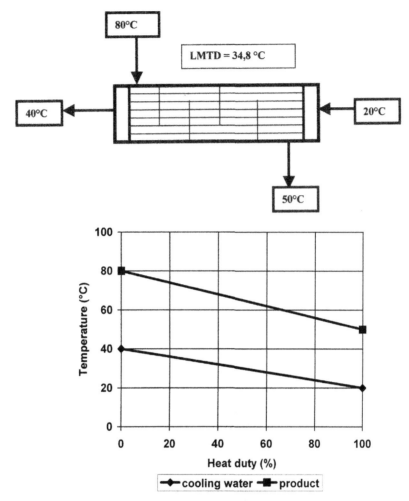

Figure 2.2 Temperature profile as function of the heat load in a one-pass heat exchanger with ideal countercurrent flow.

The temperature efficiency factor F can be calculated for a heat exchanger shell type E according to TEMA using the following equation:

$$F = \left(\frac{\sqrt{R^2+1}}{R-1}\right) \frac{\ln[(1-P_z)/(1-RP_z)]}{\ln\left[\dfrac{(2/P_z)-1-R+\sqrt{R^2+1}}{(2/P_z)-1-R-\sqrt{R^2+1}}\right]}$$

$$P_z = \frac{1-\left(\dfrac{RP-1}{P-1}\right)^{1/N}}{R-\left(\dfrac{RP-1}{P-1}\right)^{1/N}}$$

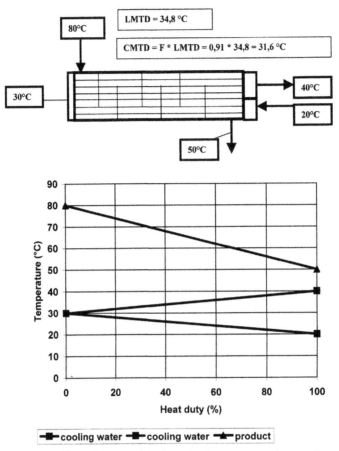

Figure 2.3 Temperature profile as function of the heat load in a two-pass heat exchanger with co- and countercurrent flow.

$$P = \frac{t_2 - t_1}{T_1 - t_1} \quad R = \frac{T_1 - T_2}{t_2 - t_1}$$

N = number of heat exchangers in series

The following table shows how large the temperature approach or a temperature cross deteriorates the CMTD, and that one can improve the temperature efficiency factor with multiple heat exchangers in series.

T_1	T_2	t_1	t_2	N	LMTD	F	CMTD
58	42	15	35.1	1	24.9	0.906	22.6
58	42	20	40.1	1	19.9	0.844	16.8
58	42	25	45.1	1	14.9	0.670	10

T_1	T_2	t_1	t_2	N	LMTD	F	CMTD
58	42	25	45.1	2	14.9	0.936	13.9
58	42	30	50.1	2	9.8	0.840	8.2
58	42	35	55.1	4	4.7	0.818	3.8
58	42	37	57.1	6	2.4	0.631	1.5
58	42	37	57.1	7	2.4	0.764	1.8

Corollary:

1. With large approach of temperatures at the tube and shell side, the temperature efficiency factor F and hence the CMTD deteriorates.
2. Through the arrangement of multiple heat exchangers one after the other, one can improve the temperature efficiency factor F because with this arrangement one approaches the countercurrent flow.
3. A large temperature cross is only possible with multiple equipments in series.

The *temperature efficiency factor F* can also be determined graphically using the diagrams in Figure 2.4 with the calculation variables P and R.

Alternatively, the *CMTD* in multipass heat exchangers can be determined with the calculation variables O and M.

$$\text{CMTD} = \frac{M}{\ln\dfrac{O+M}{O-M}} \, (^{\circ}\text{C})$$

$$O = (T_1 - t_2) + (T_2 - t_1)$$

$$M = \sqrt{(T_1 - T_2)^2 + (t_2 - t_1)^2}$$

Example 2: Calculation of LMTD and CMTD for the heat exchangers in Figures 2.2 and 2.3

Shell inlet temperature $T_1 = 80\,^{\circ}\text{C}$
Shell outlet temperature $T_2 = 50\,^{\circ}\text{C}$
Tube inlet temperature $t_1 = 20\,^{\circ}\text{C}$
Tube outlet temperature $t_2 = 40\,^{\circ}\text{C}$
$\Delta t_1 = T_1 - t_2 = 80 - 40 = 40\,^{\circ}\text{C}$
$\Delta t_2 = T_2 - t_1 = 50 - 20 = 30\,^{\circ}\text{C}$
Calculation of the LMTD for ideal countercurrent in Figure 2.1:

$$\text{LMTD} = \frac{40 - 30}{\ln\frac{40}{30}} = 34.8\,^{\circ}\text{C}$$

Determination of the CMTD for the two-pass heat exchanger in Figure 2.2.

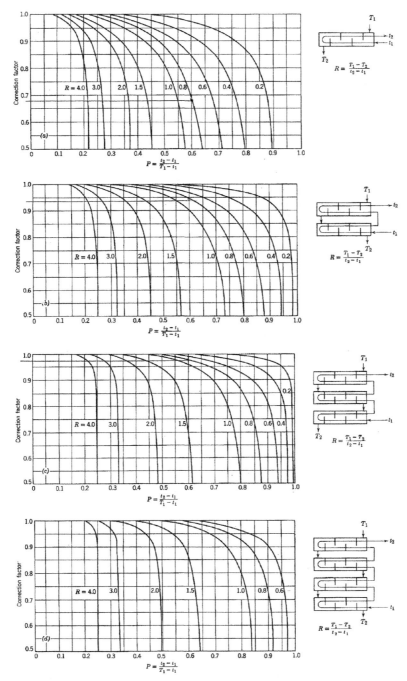

Figure 2.4 Diagrams with the temperature correction factors as function of *P* and *R*. *(From Fraas, Ozisik, Heat Exchanger Design, Wiley, 1965 [9].)*

Determination of the calculation variables P and R:

$$P = \frac{t_2 - t_1}{T_1 - t_1} = \frac{40 - 20}{80 - 20} = 0.333 \quad R = \frac{T_1 - T_2}{t_2 - t_1} = \frac{80 - 50}{40 - 20} = 1.5$$

For the variables P and R, one takes from the above diagram in Figure 2.4, the efficiency temperature factor $F = 0.91$.

Calculation of the CMTD for nonideal countercurrent:

$$CMTD = F \times LMTD = 0.91 \times 34.8 = 31.6\,°C$$

In the two-pass heat exchanger with nonideal countercurrent flow, the effective mean temperature difference of $LMTD = 34.8\,°C$ deteriorates to $CMTD = 31.6\,°C$.

Alternative calculation of CMTD with M and O:

$$O = (80 - 40) + (50 - 20) = 70$$

$$M = \sqrt{(80 - 50)^2 + (40 - 20)^2} = 36$$

$$CMTD = \frac{36}{\ln\dfrac{70 + 36}{70 - 36}} = 31.6\,°C$$

Example 3: Correction factor F for the arrangement of heat exchangers in series

$$T_1 = 58\,°C \quad T_2 = 42\,°C \quad t_1 = 25\,°C \quad t_2 = 45.1\,°C$$

$$P = \frac{t_2 - t_1}{T_1 - t_1} = \frac{45.1 - 25}{58 - 25} = 0.6 \quad R = \frac{T_1 - T_2}{t_2 - t_1} = \frac{58 - 42}{45.1 - 25} = 0.8$$

The correction factor F follows from the intersection of $P = 0.6$ on the x-axis with the curve for $R = 0.8$ in Figure 2.4.

For a single heat exchanger	$F = 0.67$
Two exchangers in series	$F = 0.94$
Three exchangers in series	$F = 0.97$
Four exchangers in series	$F = 1$

Through the arrangement of heat exchangers in series, the countercurrent effect is improved and therefore achieves a better usage of the temperature difference.

Example 4: Calculation of the temperature factor F and the CMTD for four heat exchangers in series

$$T_1 = 58\,°C \quad T_2 = 42\,°C \quad t_1 = 35\,°C \quad t_2 = 55.1\,°C$$

$$\Delta t_1 = 58 - 55.1 = 2.9\,°C \quad \Delta t_2 = 42 - 35 = 7\,°C$$

$$\text{LMTD} = \frac{7 - 2.9}{\ln\frac{7}{2.9}} = 4.65\,°C \quad R = \frac{58 - 42}{55.1 - 35} = 0.796 \quad P = \frac{55.1 - 35}{58 - 35} = 0.8739$$

$F = 0.8186$ for four heat exchangers in series from Figure 2.4.

With the temperature efficiency factor F, the CMTD is determined as follows:

$$\text{CMTD} = 0.8186 \times 4.65 = 3.8\,°C$$

Example 5: Calculation of the CMTD for isothermal and nonisothermal heating

Tube side: $t_1 = 30\,°C$ $t_2 = 60\,°C$

Shell side: Isothermal heating ($F = 1$)   Nonisothermal heating ($F < 1$)

$T_1 = T_2 = 100\,°C$ $T_1 = 100\,°C \quad T_2 = 80\,°C$

$\Delta t_1 = 100 - 60 = 40\,°C$ $\Delta t_1 = 100 - 60 = 40\,°C$

$\Delta t_2 = 100 - 30 = 70\,°C$ $\Delta t_2 = 80 - 30 = 50\,°C$

$$\text{LMTD} = \frac{70 - 40}{\ln\frac{70}{40}} = 53.6\,°C \qquad\qquad \text{LMTD} = \frac{50 - 40}{\ln\frac{50}{40}} = 44.8\,°C$$

$F = 1$ with isothermal heating $F = 0.948$

$\text{CMTD} = 1 \times 53.6 = 53.6\,°C$ $\text{CMTD} = 0.948 \times 44.8 = 42.5\,°C$

Alternative calculation for CMTD with isothermal heating:

$$O = (100 - 60) + (100 - 30) = 110$$

$$M = \sqrt{(100 - 100)^2 + (60 - 30)^2} = 30$$

$$\text{CMTD} = \frac{30}{\ln\frac{110 + 30}{110 - 30}} = 53.6\,°C$$

Alternative calculation for CMTD with nonisothermal heating:

$$O = (100 - 60) + (80 - 30) = 90$$

$$M = \sqrt{(100 - 80)^2 + (60 - 30)^2} = 36$$

$$\text{CMTD} = \frac{36}{\ln\frac{90 + 36}{90 - 36}} = 42.5\,°C$$

In Figure 2.5, it is shown how to connect multiple heat exchangers in series.

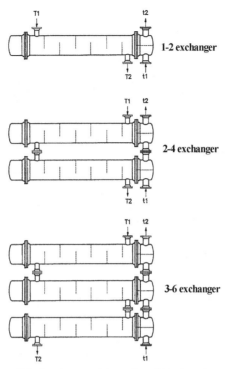

Figure 2.5 Arrangement in series of heat exchangers.

2.3 INFLUENCE OF BYPASS STREAMS ON LMTD

An important issue that has not been considered so far is the temperature profile distortion.

On the shell side not the total product stream flows across through the tube bundle and becomes heated or cooled. A part of the product stream flows in bypass or through leakage away from the heat exchanger tubes and does not take part in heat exchange.

Therefore, only a part of the shell–side product stream is heated or cooled and this part is heated or cooled more because the flow is smaller.

Hence the driving temperature gradient is reduced.

The LMTD becomes smaller with increasing leakage.

This is shown in Figure 2.6 for the following conditions:

Heating medium: Steam at 100 °C, isotherm

Product heating from 70 °C to 90 °C

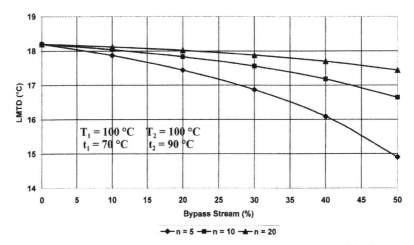

Figure 2.6 Logarithmic mean temperature difference (LMTD) as a function of the bypass leakage.

Inlet temperature difference $\Delta t_1 = T_1 - t_2 = 100 - 70 = 30\,°C$
Outlet temperature difference $\Delta t_2 = T_2 - t_1 = 100 - 90 = 10\,°C$

$$\text{LMTD} = \frac{30 - 10}{\ln \frac{30}{10}} = 18.2\,°C \text{ at 0\% bypass}$$

Figure 2.6 shows that the increasing number n of change in reversing the flow direction the temperature efficiency becomes better. This is because with each shell-side pass, both streams mix.

Recommendation: Baffle spacing $= 0.2 \times$ shell diameter or maximum $2\,°C$ temperature change per shell pass.

Figure 2.7 shows the deficiency in the mean temperature difference with increasing leakage.

Figure 2.8 shows how the logarithmic differential temperature difference (LDT) between the hot and the cold stream in the individual zones along the length of the heat exchanger decreases. The entering cold product has a high driving temperature difference LDT which decreases with progressing product heating over the length of the heat exchanger. The ideal LDT line without leakage and the LDT line for 50% bypass flow are shown. The driving temperature difference is considerably reduced by the bypass leakage.

2.4 MEAN WEIGHTED TEMPERATURE DIFFERENCE

For nonstraight line temperature curves in the temperature–heat load diagram, for instance in multicomponents condensation, large heat loads are required at high-temperature gradients and smaller heat loads with low-temperature differences.

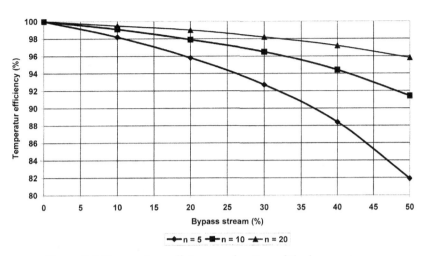

Figure 2.7 Temperature efficiency as function of the bypass stream.

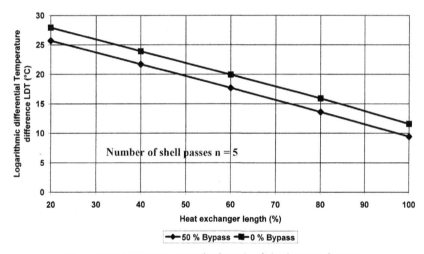

Figure 2.8 LDT curve over the length of the heat exchanger.

A normal calculation of the driving temperature difference CMTD is wrong.

The CMTDs in the different load zones are determined and lastly, the WMTD is calculated.

An example is shown in Figure 2.9.

The difference between CMTD = 67.32 K and WMTD = 103.78 is considerable.

The condenser would have been oversized with CMTD = 67.38 °C by 54%.

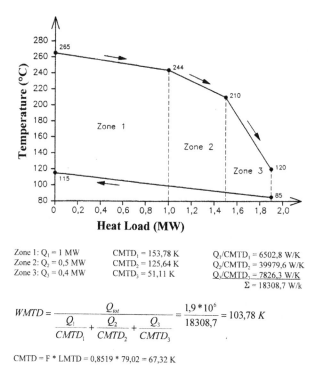

Zone 1: $Q_1 = 1$ MW $CMTD_1 = 153,78$ K $Q_1/CMTD_1 = 6502,8$ W/K
Zone 2: $Q_2 = 0,5$ MW $CMTD_2 = 125,64$ K $Q_2/CMTD_2 = 39979,6$ W/K
Zone 3: $Q_3 = 0,4$ MW $CMTD_3 = 51,11$ K $Q_3/CMTD_3 = 7826,3$ W/K
                                                           $\Sigma = 18308,7$ W/k

$$WMTD = \frac{Q_{tot}}{\dfrac{Q_1}{CMTD_1} + \dfrac{Q_2}{CMTD_2} + \dfrac{Q_3}{CMTD_3}} = \frac{1,9 * 10^6}{18308,7} = 103,78 \ K$$

$CMTD = F * LMTD = 0,8519 * 79,02 = 67,32$ K

Figure 2.9 Example for the determination of the mean weighted temperature difference (WMTD).

2.5 DETERMINATION OF THE HEAT EXCHANGER OUTLET TEMPERATURES

In a normal case, the required heat exchanger area for a specified heating or cooling, that is, for a specified temperature difference, is determined.

In this chapter, it will be shown how to calculate the outlet temperatures of a specified heat exchanger for different product stream rates.

Required data:

W_h = hot stream flow rate (kg/h)

t_{h1} = inlet temperature of the hot stream (°C)

t_{h2} = outlet temperature of the hot stream (°C)

c_h = specific heat capacity of the hot medium (Wh/kg K)

W_c = cold stream flow rate (kg/h)

T_{c1} = inlet temperature of the cold medium (°C)

T_{c2} = outlet temperature of the cold medium (°C)

C_c = specific heat capacity of the cold medium (Wh/kg K)

A = heat exchanger area (m^2)

U = overall heat transfer coefficient (W/m^2 K)

F = temperature efficiency factor for nonideal countercurrent

2.5.1 Calculation of the outlet temperatures in a multipass heat exchanger under consideration of the temperature efficiency factor F for nonideal countercurrent flow

$$t_{h2} = \frac{(1 - R) \times t_{h1} + (1 - B_2) \times R \times T_{c1}}{1 - (R \times B_2)} \ (^{\circ}C)$$

$$R = \frac{W_c \times c_c}{W \times c_h}$$

$$B_2 = \exp\left[\frac{F \times U \times A}{W_c \times c_c} \times (R - 1)\right]$$

2.5.1.1 Calculation of the outlet temperature T_{c2} of the cold medium

$$T_{c2} = \frac{t_{h1} - t_{h2}}{R} + T_{c1} \ (^{\circ}C)$$

2.5.2 Calculation of the outlet temperature t_{h2} for ideal countercurrent without F

$$t_{h2} = \frac{(1 - R) \times t_{h1} + (1 - B_1) \times R \times T_{c1}}{1 - (R \times B_1)} \ (^{\circ}C)$$

$$B_1 = \exp\left[\frac{U \times A}{W_c \times c_c} \times (R - 1)\right]$$

Example 6: Calculation of the heat exchanger outlet temperatures t_{h2} and T_{c2}

Hot side: W_h = 10 t/h t_{h1} = 58 °C t_{h2} = 42 °C c_h = 0.5 Wh/kg K
Cooling heat load $Q = 10000 \times 0.5 \times (58 - 42) = 80{,}000$ W
Cold side: cooling water T_{c1} = 25 °C T_{c2} = 45.1 °C c_c = 1.16 Wh/kg K

Overall heat transfer coefficient U = 400 W/m² K
Temperature efficiency factor F = 0.67

Determination of the required cooling water rate for the cooling heat load Q = 80 kW

$$W_c = \frac{Q}{c_c \times (T_{c2} - T_{c1})} = \frac{80000}{1.16 \times (45.1 - 25)} = 3431 \text{ kg/h}$$

Calculation of LMTD and CMTD

$$
\begin{array}{ccc}
58 & \rightarrow & 42 \\
\underline{45.1} & \leftarrow & \underline{25} \\
12.9 & & 17
\end{array}
$$

$$\text{LMTD} = \frac{17 - 12.9}{\ln \frac{17}{12.9}} = 14.9\,°C$$

$$P = \frac{45.1 - 25}{58 - 25} = 0.609 \quad R = \frac{58 - 42}{45.1 - 25} = 0.796 \quad F = 0.67$$

Calculation of the heat exchanger area A:

$$A = \frac{Q}{U \times F \times \text{LMTD}} = \frac{80000}{400 \times 0.67 \times 14.9} = 20\,m^2$$

Calculation of the auxiliary parameters R and B_2:

$$R = \frac{W_c \times c_c}{W_h \times c_h} = \frac{3431 \times 1.16}{10000 \times 0.5} = 0.796$$

$$B_2 = \exp\left[\frac{F \times U \times A}{W_c \times c_c} \times (R - 1)\right] = \exp\left(\frac{0.67 \times 400 \times 20}{3431 \times 1.16} \times (0.796 - 1)\right) = 0.7598$$

Calculation of the outlet temperature t_{h2} of the hot medium:

$$t_{h2} = \frac{(1 - R) \times t_{h1} + (1 - B_2) \times R \times T_{c1}}{1 - (R \times B_2)}$$

$$t_{h2} = \frac{(1 - 0.796) \times 58 + (1 - 0.7598) \times 0.796 \times 25}{1 - (0.796 \times 0.7598)} = 42\,°C$$

Calculation of the outlet temperature T_{c2} of the cold stream

$$T_{c2} = \frac{t_{h1} - t_{h2}}{R} + T_{c1} = \frac{58 - 42}{0.796} + 25 = 45.1\,°C$$

Example 7: Calculation of the cooling water outlet temperature T_{c2} for a smaller cooling water flow rate

$$W_c = 3135\,kg/h \text{ instead of } 3431\,kg/h \qquad \text{Data: see example 6}$$

The *cooling water outlet temperature* rises from 42 °C to 47 °C by reducing the cooling water flow rate from 3431 kg/h to 3135 kg/h.

Thereby reducing the CMTD from 9.95 °C to CMTD = 6.42.

$$\text{LMTD} = 13.78\,°C \quad R = 0.727 \quad P = 0.666 \quad F = 0.465 \quad \text{CMTD} = 6.42\,°C$$

The lower effective temperature difference reduces the heat duty within the heat exchanger from 80 kW for CMTD = 9.95—51.36 kW for CMTD = 6.42 °C:

$$Q = k \times A \times \text{CMTD} = 400 \times 20 \times 6.42 = 51360\,W$$

In addition, it has to be considered that through the smaller product rate, the flow velocity and hence also the heat transfer coefficient will be reduced and so that the overall heat transfer coefficient U becomes smaller.

REFERENCES AND FURTHER READING

[1] R.A. Bowman, Ind. Engr. Chem. 28 (1936), 541/544.
[2] R.A. Bowman, Trans. ASME 62 (1940), 283/294.
[3] J. Fernandez, Trans. ASME (1957), 287/297.
[4] F.K. Fisher, Ind. Engr. Chem. 30 (1938), 377/383.
[5] J. Bowman, R. Turton, Chem. Eng (Juli 1990).
[6] Ch.A. Plants, Chem. Eng. (Juni 1992 + Juli 1992).
[7] R. Mukherjes, Chem. Eng. Prog. (April 1996 + Februar 1998).
[8] J. Fisher, R.O. Parker, Hydrocarbon Process. (Juli 1969).
[9] A.P. Fraas, M.N. Ozisik, Heat Exchanger Design, John Wiley, N.Y., 1965.

CHAPTER 3

Calculations of the Heat Transfer Coefficients and Pressure Losses in Convective Heat Transfer

Contents

How are the heat transfer coefficients calculated? [1,3–5,11,14].

The required calculation equations for the determination of the heat transfer coefficients on the tube side and on the shell side are listed in Figure 3.2. The application of the equations is shown in Examples 1 and 2. The calculation of the heat transfer coefficients is simple and the results are on the safe side.

The relationships are valid for the following geometrical conditions:

Baffle spacing/diameter $B/D = 0.2–0.3$

Segmental height/diameter $H/D = 0.2–0.3$

These geometrical conditions have proved to be effective in practice.

The calculation equations in Figure 3.2 indicate that the Nusselt number Nu, that is, the heat transfer coefficient α is dependent on the Prandtl and the Reynolds number.

The Prandtl number Pr contains only the physical properties that cannot be changed. For the design in general, the properties in the heat exchanger at average temperature are chosen.

If, however, the properties in the heat exchanger greatly change with temperature a segmental calculation of the individual zones is recommended.

Heat Exchanger Design Guide
http://dx.doi.org/10.1016/B978-0-12-803764-5.00003-1

© 2016 Elsevier Inc.
All rights reserved.

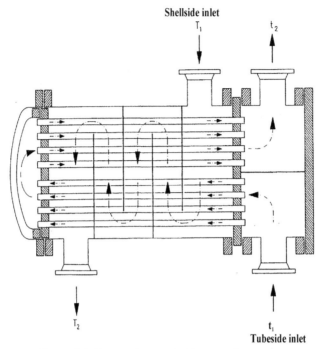

Figure 3.1 Heat exchangers with two tube passes.

The improvement of the heat transfer coefficients can only be achieved by increasing the Reynolds number Re, that is, by increasing the flow velocity w.

With increasing Reynolds number the laminar thickness of the boundary layer on the tube wall decreases and the quotient $\alpha \approx \lambda/s$ becomes higher. The following dependencies of the heat transfer coefficient α on the flow velocity w follow from the calculation equations in Figure 3.2:

Tube side: $\alpha \propto w^{0.8}$ *Shell side*: $\alpha \propto w^{0.6}$

A sufficiently high flow velocity is the required condition for a good heat transfer coefficient.

The recommended flow velocities are 0.5–2 m/s for liquids and 10–25 m/s for gases. With viscous media the shell-side heat transfer coefficients are higher than the α-values on the tube side. This is why viscous products are preferably located at the shell side.

Tube-side heat transfer coefficient according to Nitsche method:

Laminar region with Re < 2300:

$$Nu = 1.86 \times \left(\frac{Re \times Pr \times d_i}{l} \right)^{0.33}$$

Transition region with 2300 < Re < 8000

$$Nu = \left(0.037 \times Re^{0.75} - 6.66 \right) \times Pr^{0.42}$$
$$0.6 < Pr < 500$$

Turbulence region with Re > 8000

$$Nu = 0.023 \times Re^{0.8} \times Pr^{0.33}$$

$$\alpha = \frac{Nu \times \lambda}{d_i} \quad Re = \frac{w \times d_i}{v} \quad Pr = \frac{3600 \times v \times c \times \rho}{\lambda} \quad t_m = \frac{t_1 + t_2}{2}$$

Shell-side heat transfer coefficient according to Nitsche method for Re > 10

Triangle pattern	**Quadratic pattern**
$Nu = 0.196 \times Re^{0.6} \times Pr^{0.33}$	$Nu = 0.156 \times Re^{0.6} \times Pr^{0.33}$

$$\alpha = \frac{Nu \times \lambda}{d_o} \qquad Re = \frac{w \times d_o}{v} \qquad w_{cross} = \frac{V (m^3/h)}{B \times (D_i - n_{acr} \times d_o)}$$

Figure 3.2 Calculation equations for the convective heat transfer.

Example 1: Calculation of the tube-side heat transfer coefficient

Basic data:

Heat exchanger DN 500 with 164 tubes $d_i = 21$ mm $n_P =$ four tube passes
25 × 2
41 tubes per pass
Tube-side throughput $V_T = 51.12$ m³/h
Density $\rho = 995$ kg/m³   Specific heat capacity $c = 1.16$ Wh/kg K
Kinematic viscosity $v = 0.92$ mm²/s   Heat conductivity $\lambda = 0.605$ W/m K

1. Calculation of the flow cross section a_T of 41 tubes

$$a_T = \frac{n}{n_P} \times d_i^2 \times \frac{\pi}{4} = \frac{164}{4} \times 0.021^2 \times 0.785 = 0.0142 \text{ m}^2$$

2. Determination of the flow velocity w_T in the tubes

$$w_T = \frac{V\left(m^3/h\right)}{a_T \times 3600} = \frac{51}{0.0142 \times 3600} = 1\ m/s$$

3. Determination of the Reynolds number

$$Re = \frac{w_T \times d_i}{\nu} = \frac{1 \times 0.021}{0.92 \times 10^{-6}} = 22826$$

4. Calculation of the Pr number

$$Pr = \frac{\nu \times c \times \rho \times 3600}{\lambda} = \frac{0.92 \times 10^{-6} \times 1.16 \times 995 \times 3600}{0.605} = 6.32$$

5. Determination of the Nusselt number

$$Nu = 0.023 \times Re^{0.8} \times Pr^{0.33} = 0.023 \times 22826^{0.8} \times 6.32^{0.33} = 129.62$$

6. Calculation of the heat transfer coefficient

$$\alpha = \frac{Nu \times \lambda}{d_i} = \frac{129.62 \times 0.605}{0.021} = 3734\ W/m^2\ K$$

Example 2: Calculation of the shell-side heat transfer coefficient

Basic data:

Shell diameter $D_i = 1.1$ m	270 tubes 38 × 2 mm	$d_o = 38$ mm
Triangular pitch $T = 50$ mm	Baffle spacing $B = 200$ mm	
Flow throughput $\overset{''}{V}_{Shell} = 200$ m^3/h		
Density $\rho = 803$ kg/m^3	Specific heat capacity $c = 2.59$ kJ/kg K = 0.721 Wh/kg K	
Kinematic viscosity $\nu = 051$ mm^2/s	Heat conductivity $\lambda = 0.108$ W/m K	

1. First, the flow cross section a_{cross} for the cross stream with the pitch T is determined

$$a_{cross} = D_i \times B \times \left(1 - \frac{d_o}{T}\right) = 1.1 \times 0.2 \times \left(1 - \frac{38}{50}\right) = 0.0528\ m^2$$

Alternative calculation with the number of tubes n_{acr} in the cross stream

$$n_{acr} = \frac{D_i}{T} = \frac{1100}{50} = 22$$

$$a_{cross} = B \times (D_i - n_{acr} \times d_o) = 0.2 \times (1.1 - 22 \times 0.038) = 0.0528\ m^2$$

2. Thereafter, the flow velocity w_{cross} for the cross stream is determined

$$w_{cross} = \frac{V_M\left(m^3/h\right)}{a_{cross} \times 3600} = \frac{200}{0.0528 \times 3600} = 1.052\ m/s$$

3. Calculation of the Reynolds number

$$Re = \frac{w_{cross} \times d_o}{\nu} = \frac{1.052 \times 0.038}{0.51 \times 10^{-6}} = 78,398$$

4. Calculation of the Pr number

$$Pr = \frac{\nu \times c \times \rho \times 3600}{\lambda} = \frac{0.51 \times 10^{-6} \times 0.721 \times 803 \times 3600}{0.108} = 9.84$$

5. Determination of the Nusselt number for triangular pitch

$$Nu = 0.196 \times Re^{0.6} \times Pr^{0.33} = 0.196 \times 78398^{0.6} \times 9.84^{0.33} = 360.18$$

6. Calculation of the heat transfer coefficient

$$\alpha = \frac{Nu \times \lambda}{d_o} = \frac{360.18 \times 0.108}{0.038} = 1023.7 \, W/m^2 \, K$$

3.1 TUBE-SIDE HEAT TRANSFER COEFFICIENT [3,6,12,14,16]

The flow velocity and the heat transfer coefficient on the tube side can be increased through a multipass arrangement. The medium flows thereby in multiple passes through the tubes and the tube-side flow cross section is reduced. In Figure 3.1 a tube bundle with two tube-side passes is depicted. With two passes the tube-side flow cross section is halved and therefore the flow velocity is doubled. The higher flow velocity in the tube improves the heat transfer coefficient. In Figure 3.3 the dependency of the heat transfer coefficient on the flow velocity and on the tube diameter is shown.

A disadvantage of the multipass arrangement is the reduction of the effective temperature difference for the heat transfer because there is no ideal countercurrent flow.

In the two passes on the tube side the medium flows in the first pass in counter-current flow and in the second pass in cocurrent flow to the product stream on the shell side.

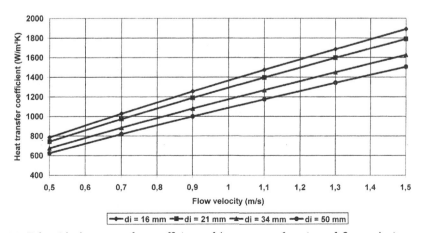

Figure 3.3 Tube-side heat transfer coefficient of heptane as function of flow velocity and tube diameter.

Therefore, the logarithmic mean temperature difference (LMTD) for countercurrent flow must be corrected with the temperature effective factor F.

$$\text{CMTD} = F \times \text{LMTD}$$

CMTD is the corrected effective mean temperature difference for the heat transfer. The calculation of CMTD is shown in Chapter 2.

An additional disadvantage of the multipass arrangement is the increase of the pressure loss through the multiple flow redirection and because of the higher flow velocity which goes into the pressure loss calculation in a square function.

In Figures 3.4 and 3.5 the heat transfer coefficients for some liquids and gases are shown.

3.2 SHELL-SIDE HEAT TRANSFER COEFFICIENT [1,2,4,9,10–12,14–16]

With the convective heat exchanger shown in Figure 3.1, the medium flows on the shell side across the tube bundle and lengthways through the baffle cuts. The flow cross-

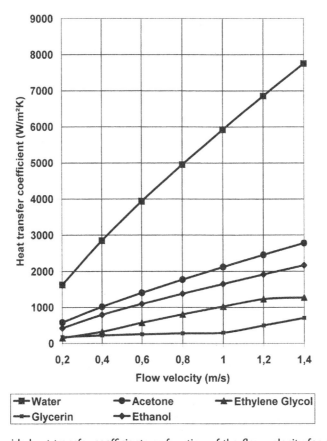

Figure 3.4 Tube-side heat transfer coefficients as function of the flow velocity for different liquids.

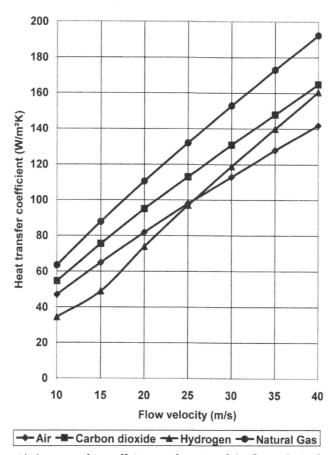

Figure 3.5 Tube-side heat transfer coefficients as function of the flow velocity for different gases.

sectional areas should be preferably equal for the cross stream through the bundle and the lengthways stream in the baffle cuts. In this case, there are equal flow velocities without acceleration turbulences with associated pressure losses.

With increasing flow velocity, the heat transfer coefficient α becomes better.

The shell–side stream velocity for the cross flow can be calculated with the following equations:

$$w_{\mathrm{cross}} = \frac{V\left(\mathrm{m}^3/\mathrm{h}\right)}{3600 \times B \times \left(D_i - n_{\mathrm{acr}} \times d_o\right)}\left(\mathrm{m/s}\right)$$

$$w_{\mathrm{cross}} = \frac{V\left(\mathrm{m}^3/\mathrm{h}\right)}{3600 \times D_i \times B \times \left(1 - \frac{d_o}{T}\right)}\left(\mathrm{m/s}\right)$$

By decrease of the baffle spacing B the flow velocity is increased and the heat transfer coefficient becomes better.

In Figures 3.6 and 3.7 it is shown how the shell-side heat transfer coefficient of liquids and gases rises with increasing flow velocity.

Through the arrangement of more baffles to improve the flow velocity, the pressure loss in the cross stream across the tube bundle is increased so that the product flow increases and hence the product flow rate flowing through the leak and bypass cross section aside the tube bundle and does not take part in heat change.

This deteriorates the heat transfer.

Figure 3.8 shows that according to VDI-Wärmeatlas [11] the correction factor f_W for the shell-side heat transfer coefficient becomes smaller with decreasing B/D ratio, that is, increasing number of baffles because each baffle causes additional bypass and leaking streams.

The correction factor f_W is defined as follows:

$$f_W = f_G \times f_L \times f_B$$

f_G is the geometrical factor which essentially considers how many heat exchanger tubes lie in the baffle window.

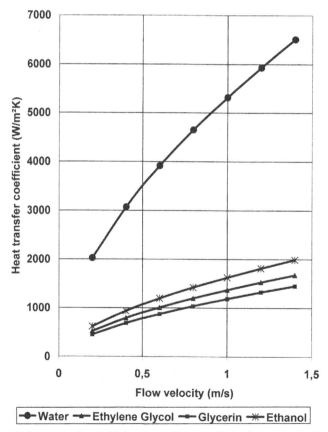

Figure 3.6 Shell-side heat transfer coefficients as a function of the flow velocity for different liquids.

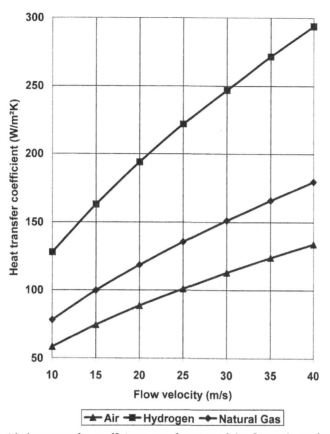

Figure 3.7 Shell-side heat transfer coefficients as a function of the flow velocity for different gases.

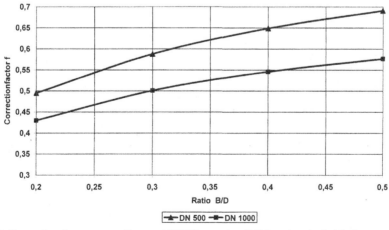

Figure 3.8 Correction factor according to VDI-Wärmeatlas [11] for the shell-side heat transfer as a function of B/D ratio. $B =$ baffle spacing, $D =$ shell diameter.

f_L is the leakage factor for the leaking stream through the gap between the tube and the baffle holes and also between the baffles and the shells.

The f_L factor is dependent on the specified tolerance and the number of the baffles. f_B is the bypass factor for the flow through the tube lanes and the annular gap between the tube bundle and the shell. With sealing strips and dummy tubes, the bypass stream can be minimized such that $f_B = 1$.

In Figure 3.9 the different streams on the shell side of a heat exchanger according to Tinker [9] are shown.

Stream A: leakage stream through the gap between the tubes and the baffle holes
Stream B: cross stream through the bundle
Stream C: bypass stream through the annular gap between the bundle and the shell
Stream E: bypass stream between the baffle and the shell
Stream F: bypass stream due to the nature of the construction caused tube lanes

In Figure 3.10 it is shown for a DN 500 heat exchanger that stream B, that is, the flow part flowing across the tube bundle increases with increasing baffle spacing. Less baffles result in the occurring of less leakage and bypass streams but less baffles with wider baffle spacings decrease the flow velocity and deteriorate the heat transfer.

These curves are valid for the following heat exchanger data:
Shell diameter DN 500 with 256 tubes 20 × 2, 6 m long, triangular pitch 26 mm, four passes.
Shell-side product: 15 m³/h kerosine, which is cooled with water from 114 to 40 °C.

From Figure 3.10 the influence of the pass partition of a multipass equipment on the bypass stream can be seen. A ribbon pass arrangement across the flow gives more cross flow than a pass arrangement in quadrants because the tube lanes in flow direction in a quadrant partition permit a higher stream F through the tube lanes.

In Figure 3.11 it can be seen that the heat transfer coefficients rise with reducing baffle spacing in spite of the increase in bypass and leakage streams. The values are valid for a constant flow throughput of 15 m³/h.

The improvement due to the higher flow velocity is greater than the deterioration by the increasing bypass and leakage streams.

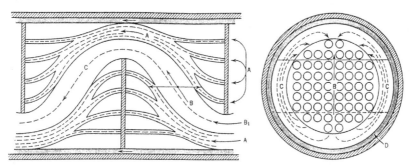

Figure 3.9 Shell-side product streams according to Tinker [3,9].

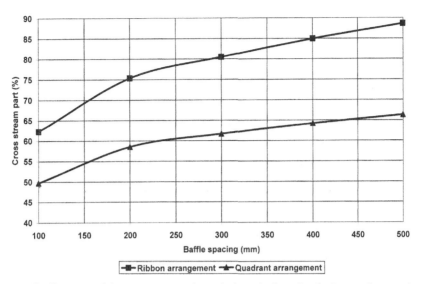

Figure 3.10 The flow part of the cross stream through the tube bundle of a DN 500 heat exchanger as function of the baffle spacing with ribbon and quadrant arrangement.

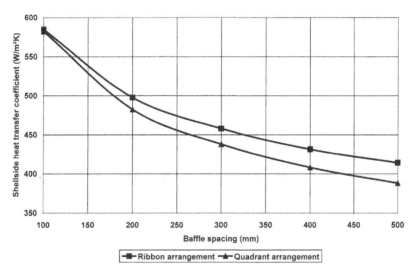

Figure 3.11 Shell-side heat transfer coefficients as function of baffle spacing for ribbon and quadrant arrangement.

The leakage and bypass streams also worsen the temperature efficiency for the heat exchange because a part of the product stream does not take part in the heat exchange process and therefore reduces the effective driving temperature difference for the heat exchange.

The reduced product rate without the bypass and leakage stream part is further heated and therefore reduces the temperature difference between the product and the heating medium.

In each pass through the baffle chamber a mixed temperature of the cross stream and the bypass/leakage stream results.

In Figure 3.12 different baffle types are shown.

In practice the segmental baffles with horizontal chord shown below are preferably used for the convective heat transfer because the medium is better mixed by the up and down streams and possible temperature differences are compensated.

With condensers vertical baffle, cuts with vertical chord are advantageous in order to avoid condensate built-up and allow the noncondensable gases to escape and flow out upward.

If the pressure loss on the shell side is to be reduced disc and doughnut—or double segmental baffles are used.

The full circle baffles have the advantage that there is no bypass streams through the diametrical gap between the tube bundle and the shell; however, the mainly horizontal streams deteriorate the heat transfer coefficient.

With the double segmental baffles, the flow cross-sectional area is doubled and hence the flow velocity halved so that the pressure loss is considerably reduced at the cost of a poorer heat transfer coefficient.

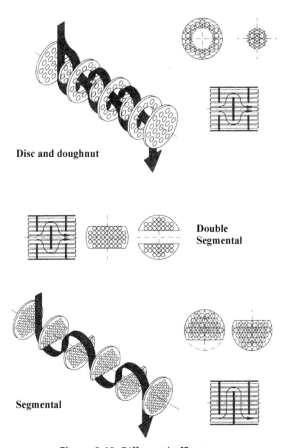

Disc and doughnut

Double Segmental

Segmental

Figure 3.12 Different baffle types.

Fundamentally the means of reducing the pressure loss on the shell side of heat exchangers are as follows:
- altering the baffle spacing and baffle arrangement
- larger pitch or quadratic pitch
- shorter tube lengths
- larger nozzle diameter
- multiple heat exchangers in parallel arrangement
- installation of a pure cross stream heat exchanger, for instance, according to TEMA J or TEMA X.

What is important for a working heat exchanger?

High flow velocity at the allowable pressure drops.

Preferably low bypass and leakage streams.

For that it is necessary:

Multiple passes on the tube side.

Small clearance tube/baffle plate holes + baffle/shell.

Small gap shell/bundle outer tube limit with small diametrical gap.

Pass arrangement in ribbons without uniformed tube lanes for the bypass.

Constructive measures:

Longitudinal plates or sealing strips in the shell in the case of large diametrical gaps or spaces between the shell and the bundle tube limit diameter.

This is especially the case for floating head equipments with large free flow cross sections between the bundle and the shell (Figure 3.13).

Dummy tubes for reducing the bypass cross sections in the tube lanes and in the diametrical gap between bundle and shell.

3.3 COMPARISON OF DIFFERENT CALCULATION MODELS

In Figures 3.14 and 3.15 the heat transfer coefficients calculated according to different models for cooling water on the tube side and ethanol on the shell side are shown.

The ethanol is cooled from 58 to 42 °C with cooling water which is heated from 15 to 35.1 °C.

The calculations were made for the heat exchanger type 10 in Table 1 in Chapter 3.6:

DN 500 with 164 tubes in four passes, tube length 3 m, baffle spacing 100 mm.

Model Nitsche method (see Figure 3.2)

Model TPLUS program: software for the design and rating of heat exchangers

Model Kern [16]

Model VDI-Wärmeatlas [11]

Corollary: From the overall heat transfer coefficients, shown in Figure 3.16, it can be seen that the deviations are not large and that one is on the safe side with the simple Nitsche method.

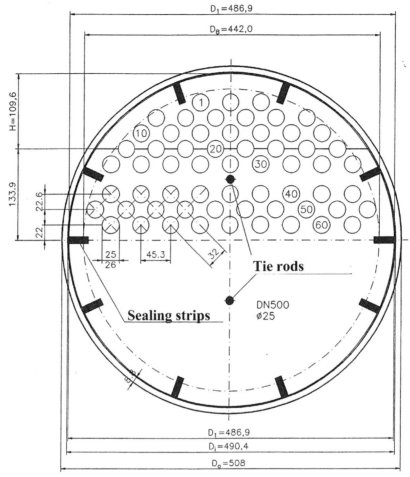

Figure 3.13 Floating head heat exchanger DN 500 with sealing strips to reduce the bypass stream and to improve the heat transfer coefficient.

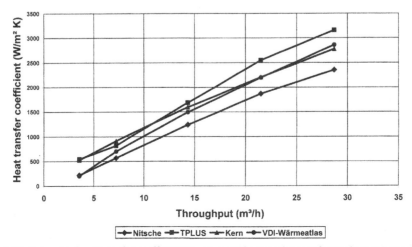

Figure 3.14 Tube-side heat transfer coefficients in a type 10 heat exchanger for cooling water in the tubes.

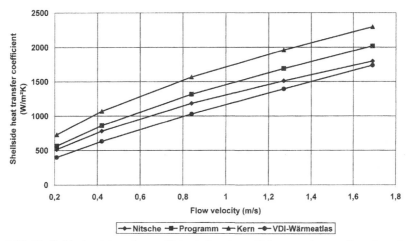

Figure 3.15 Shell-side heat transfer coefficients in a type 10 heat exchanger for ethanol at 50 °C.

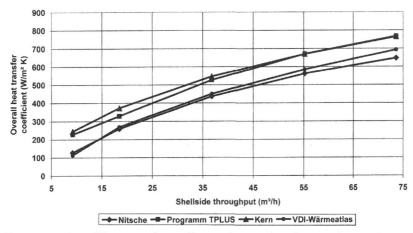

Figure 3.16 Overall heat transfer coefficients for an ethanol–water heat exchanger.

3.4 PRESSURE LOSS IN CONVECTIVE HEAT EXCHANGERS

3.4.1 Tube-side pressure drop ΔP_T

The pressure loss in the tubes is calculated as follows:

$$\Delta P_T = \frac{w_T^2 \times \rho}{2} \times n_P \times \left(\frac{f \times L}{d_i} + 4 \right) (Pa) \quad \text{Friction factor } f = \frac{0.216}{Re^{0.2}}$$

Nozzle pressure loss:

$$\text{Inlet nozzle:} \quad \Delta P_{Ni} = \frac{w_{Ni}^2 \times \rho}{2} (Pa) \quad \text{Outlet nozzle:} \quad \Delta P_{No} = 0.5 \times \frac{w_{No}^2 \times \rho}{2} (Pa)$$

ΔP_T = pipe pressure loss

ΔP_{tot} = total pressure loss = $\Delta P_T + \Delta P_{ni} + \Delta P_{no}$

w_{ni} = flow velocity in inlet nozzle (m/s)

w_{no} = flow velocity in outlet nozzle (m/s)

w_T = flow velocity in the tube (m/s)

L = tube length (m)

d_i = tube diameter (m)

n_P = number of tube passes

ρ = density (kg/m^3)

Example 3: Tube-side pressure loss

$V = 14.35$ m^3/h $\qquad w_T = 0.28$ m/s $\qquad$ Re = 6406 $\qquad L = 3$ m

$d_i = 21$ mm $\qquad n_P$ = four passes $\qquad$ 41 tubes per pass $\qquad \rho = 995$ kg/m^3

$$f = \frac{0.216}{6406^{0.2}} = 0.03742 \quad \Delta P_T = \frac{0.28^2 \times 995}{2} \times 4 \times \left(\frac{0.03742 \times 3}{0.021} + 4\right) = 1458 \text{ Pa}$$

Inlet nozzle: DN100 $\quad w_{Ni} = 0.5077$ m/s $\quad \Delta P_{Ni} = \dfrac{0.5077^2 \times 995}{2} = 128$ Pa

Outlet nozzle: DN80 $\quad w_{No} = 0.793$ m/s $\quad \Delta P_{No} = 0.5 \times \dfrac{0.793^2 \times 995}{2} = 156$ Pa

$\Delta P_{tot} = 1458 + 128 + 156 = 1742$ Pa

In Figure 3.17 the calculated pressure losses using the different models for water in a heat exchanger type 10 in Table 1.1 (Chapter 1) are shown.

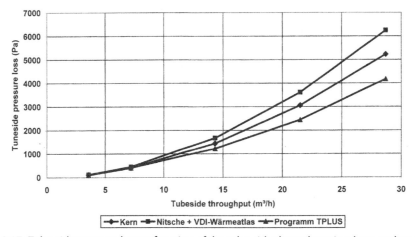

Figure 3.17 Tube-side pressure loss as function of the tube-side throughput in a heat exchanger type 10 according to Table 1.1 in Chapter 1.

3.4.2 Shell-side pressure loss

The total pressure loss ΔP_{tot} on the shell side is made of
- the pressure loss ΔP_W in the flow through the baffle windows
- the pressure loss ΔP_{cr} in the cross stream through the tubes
- the nozzle pressure loss ΔP_n in inlet and outlet nozzle

$$\Delta P_{tot} = \Delta P_w + \Delta P_{cr} + \Delta P_n$$
$$\Delta P_w = BF \times n_B \times \varrho \times w_w^2$$
$$\Delta P_{cr} = BF \times (n_B + 1) \times f \times n_{cross} \times w_{cr}^2 \times \varrho/2$$

f = friction factor for cross stream = $f(Re)$
n_{cross} = tube rows in the cross stream between both baffle cuts
n_B = number of baffles w_{cr} = cross stream velocity (m/s)
ρ = density (kg/m^3) w_W = window velocity (m/s)
BF = bypass factor considering bypass and leakage streams = 0.36

Example 4: Shell-side pressure loss calculation

$V = 50$ m^3/h $\varrho = 763.4$ kg/m^3 $\nu = 0.91$ mm^2/s
$D_i = 496$ mm   164 tubes 25 × 2   Triangular pitch 32 mm
Baffle spacing $B = 100$ mm   Segment height $H = 100$ mm
Number of baffles $n_B = 28$ Flow through tube rows $n_{cross} = 9$
Flow cross section in cross stream $a_{cross} = 0.0121$ m^2 → $w_{cr} = 1.148$ m/s
Flow cross section in the windows $a_W = 0.0165$ m^2 → $w_W = 0.842$ m/s

1. Pressure loss ΔP_W in the 28 baffle windows

$$\Delta P_W = BF \times n_B \times \rho \times w_w^2 = 0.36 \times 28 \times 763.4 \times 0.842^2 = 5455 \text{ Pa}$$

The bypass factor BF = 0.36 comes from the assumption that because of the bypass streams only 60% of the geometrically calculated flow velocity will be achieved.

2. Pressure loss ΔP_{cr} with cross stream through nine tube rows

$$\Delta P_{cr} = BF \times \left((n_B + 1) \times f \times n_{cross} \times \frac{w_{cr}^2 \times \rho}{2} \right)$$

$$\Delta P_{cr} = 0.36 \times (28 + 1) \times f \times 9 \times \frac{1.148^2 \times 763.4}{2} = 47,266 \times f$$

The friction factor f is calculated according to different models with the help of the Reynolds number.

$$Re = \frac{w_{cr} \times d_o}{\nu} = \frac{1.148 \times 0.025}{0.91 \times 10^{-6}} = 31,538$$

2.1 According to Bell with f_B [10]

$$f_B = 2.68 \times Re^{-0.182} = 2.68 \times 31538^{-0.182} = 0.407$$

$$\Delta P_{crB} = 47266 \times 0.407 = 19237 \text{ Pa}$$

$$\Delta P_W + \Delta P_{cr} = 5455 + 19,237 = 24,692 \text{ Pa}$$

2.2 According to Clark–Davidson with f_{CD} [17]

$$f_{CD} = Re^{-0.2} \times \frac{3.12}{\sqrt{T/d_o}} = 31,538^{-0.2} \times \frac{3.12}{\sqrt{32/25}} = 0.347$$

$$\Delta P_{crCD} = 0.347 \times 47,266 = 16,418 \text{ Pa}$$

$$\Delta P_W + \Delta P_{cr} = 5455 + 16,418 = 21,873 \text{ Pa}$$

2.3 According to Jakob with f_J [13]

$$f_J = Re^{-0.16} \times \left(1 + \frac{0.47}{(T/d_o - 1)^{1.08}}\right) = 2.85 \times 31,538^{-0.16} = 0.545$$

$$\Delta P_{crJ} = 0.545 \times 47,266 = 25,755 \text{ Pa}$$

$$\Delta P_W + \Delta P_{cr} = 5455 + 25,755 = 31,210 \text{ Pa}$$

2.4 According to Donohue with f_D [7]

$$f_D = Re^{-0.2} \times \frac{8}{\left(\dfrac{T - d_o}{d_o}\right)^{0.2}} = 3.86 \times 31,538^{-0.2} = 0.487$$

$$\Delta P_{crD} = 0.487 \times 47,266 = 23,039 \text{ Pa}$$

$$\Delta P_W + \Delta P_{cr} = 5455 + 23,039 = 28,494 \text{ Pa}$$

2.5 According to Chopey with the bypass factor BF as a function of D_i and B [15]. Number of rows for cross stream n_{cross}:

$$n_{cross} = \frac{b \times D_i}{T} = \frac{0.7 \times 0.496}{0.032} = 10.85$$

Reynolds number and friction factor:

$$Re_{Ch} = \frac{w_{cr} \times (T - d_o)}{\nu} = \frac{1.148 \times (0.032 - 0.025)}{0.91 \times 10^{-6}} = 8830$$

$$f_{Ch} = 4 \times Re_{Ch}^{-0.25} = 0.4126$$

Pressure loss calculation:

$$\Delta P_{crCh} = BF \times (n_B + 1) \times n_{cross} \times f_{Ch} \times \frac{w_{cr} \times \rho}{2} \text{ (Pa)}$$

$$\Delta P_{crCh} = BF \times 29 \times 10.85 \times 0.4126 \times \frac{1.148^2 \times 763.4}{2} = 65312 \times BF$$

Calculation of the bypass factor BF [15] and the pressure loss:

$$BF = R_1 \times R_2$$

$$R_1 = 0.75 \times \sqrt{\frac{B}{D_i}} = 0.75 \times \sqrt{\frac{100}{496}} = 0.3368$$

$$R_2 = 0.85 \times D_i^{0.08} = 0.85 \times 0.496^{0.08} = 0.8036$$

$$BF = 0.3368 \times 0.8036 = 0.2706$$

$$\Delta P_{crCh} = 0.2706 \times 65,312 = 17,675 \text{ Pa}$$

$$\Delta P_w + \Delta P_{cr} = 5455 + 17,675 = 23,130 \text{ Pa}$$

2.6 Pressure loss according to Kern(new) with the bypass correction in f_K [16]

$$\Delta P_{MK} = (n_B + 1) \times f_K \times \frac{D_i}{d_e} \times \frac{w_K^2 \times \rho}{2} \text{ (Pa)}$$

First the special "Kern data" must be determined:

$$w_K = 1.28 \text{ m/s} \quad d_e = 0.0198 \, m \quad Re_K = 27,649$$

$$f_K = 0.436 \times \lg Re^{-1.4485} = 0.05$$

$$\Delta P_{MK} = 29 \times 0.05 \times \frac{496}{19.8} \times \frac{1.28 \times 763.4}{2} = 22,849 \text{ Pa}$$

$$\Delta P_{MK} = \Delta P_w + \Delta P_{cr} = 22,849 \text{ Pa}$$

3. Nozzle pressure loss

Flow velocity in DN 150 $\quad w_n = 0.786$ m/s

$$\Delta P_n = 1.5 \times \frac{w_n^2 \times \rho}{2} = 1.5 \times \frac{0.786^2 \times 763.4}{2} = 354 \text{ Pa}$$

4. Comparison of result for $\Delta P_{tot} = \Delta P_w + \Delta P_{cr} + \Delta P_n$

Bell	$\Delta P_{tot} = 25,046$ Pa
Clark–Davidson	$\Delta P_{tot} = 22,227$ Pa
Jakob	$\Delta P_{tot} = 31,564$ Pa
Donohue	$\Delta P_{tot} = 28,848$ Pa
Chopey	$\Delta P_{tot} = 23,484$ Pa
Kern	$\Delta P_{tot} = 23,203$ Pa

In Figure 3.18 the calculated pressure losses using the different models for ethanol in a heat exchanger type 10 with the baffle spacing of 100 mm are shown.

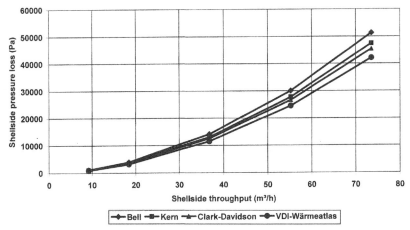

Figure 3.18 Shell-side pressure loss as function of the flow velocity in a heat exchanger type 10 according to Table 1.1 in Chapter 1.

3.5 HEAT EXCHANGER DESIGN WITH HEAT EXCHANGER TABLES

In Tables 3.1–3.4 are listed the most important data for heat exchangers according to DIN 28184 and DIN 28191.

In order to simplify the selection of the correct heat exchanger type for a specific problem definition with respect to a good flow velocity the volumetric flow rates are listed in both columns of the right side which are necessary for the flow velocity of 1 m/s:

VR = necessary flow throughput for 1 m/s on the tube side (m^3/h)

VM = necessary flow throughput for 1 m/s on the shell side (m^3/h)

The VM value on the shell side is valid for the given baffle spacing B.

The heat exchanger area per meter tube length is listed under A (m^2/m).

Procedure:

First during the design an estimated overall heat transfer coefficient for the required heat exchanger area is determined. The heat exchanger is then selected from the tables.

Criteria for the selection:

- The heat exchanger type must have an area greater than the estimated.
- The tube-side and shell-side flow rates should lie in the order of the flows VR and VM that are listed in the tables for a flow velocity of 1 m/s.

The determination of the required flow velocities for the design of the heat exchanger is very simple using the listed values of VR and VM in the table.

The application of the tables is shown in the following examples.

Table 3.1 Geometrical data of heat exchangers according to DIN 28184 part 1 (triangular pitch 60°), tubes 25 × 2, pitch 32 mm triangular, tube tolerance $\Delta d = 0.08$ mm, tube hole in the baffle $d_{hole} = 26$ mm

Type Nr.	DN	n_P	D_a mm	D_i mm	s mm	SMU mm	D_H mm	B mm	n	n_W	n_{acr}	A m²/m	f_W	VR m³/h	VM m³/h
1	150	2	168	159	4.5	1.25	143	50	14	6	4	1.09	0.3624579	8.73	8.89
2	200	2	219	207.2	5.9	1.25	191	50	26	9	6	2.04	0.4663294	16.21	9.23
3	250	2	273	260.4	6.3	1.25	248	50	44	12	6	3.45	0.3133393	27.43	18.80
4	300	2	324	309.8	7.1	1.25	298	60	66	25	9	5.18	0.4285523	41.15	17.51
5	350	2	355	339	8	1.5	316	70	76	22	10	5.96	0.4481409	47.38	21.59
6	350	4	355	339	8	1.5	325	70	68	22	10	5.34	0.4751083	21.20	21.59
7	400	2	406	388.4	8.8	1.5	373	80	106	35	10	8.32	0.3562161	66.09	38.90
8	400	4	406	388.4	8.8	1.5	369	80	88	26	11	6.91	0.4420558	27.43	31.79
9	500	2	508	496	6	1.75	478	100	180	62	14	14.13	0.4098792	112.22	51.72
10	500	4	508	496	6	1.75	482	100	164	45	15	12.88	0.5205206	51.12	42.78
11	600	2	600	588	6	1.75	576	120	258	82	17	20.26	0.4489858	160.85	69.60
12	600	8	600	588	6	1.75	573	120	232	64	16	18.22	0.4110614	36.16	80.35
13	700	2	700	684	8	2.25	672	140	364	113	21	28.58	0.5294837	226.94	79.37
14	700	8	700	684	8	2.25	666	140	324	88	20	25.44	0.4712546	50.50	91.93
15	800	2	800	784	8	2.25	771	160	484	155	24	38.01	0.5295184	301.75	105.23
16	800	8	800	784	8	2.25	772	160	432	116	22	33.92	0.4339129	67.33	133.96
17	900	2	900	880	10	2.25	868	180	622	174	26	48.85	0.4824717	387.79	148.26
18	900	8	900	880	10	2.25	868	180	556	156	26	43.66	0.4962607	86.66	148.26
19	1000	2	1000	980	10	2.75	966	200	776	216	30	60.94	0.5353497	483.80	164.85
20	1000	8	1000	980	10	2.75	966	200	712	164	28	55.92	0.4528635	110.97	200.79
21	1100	2	1100	1076	12	2.75	1058	220	934	264	33	73.35	0.5440129	582.30	198.04
22	1100	8	1100	1076	12	2.75	1055	220	860	242	32	67.54	0.5009315	134.04	217.82
23	1200	2	1200	1176	12	2.75	1159	240	1124	328	35	88.27	0.4919161	700.76	259.30
24	1200	8	1200	1176	12	2.75	1160	240	1048	302	34	82.3	0.4633054	163.34	280.87

Table 3.2 Geometrical data of heat exchangers according to DIN 28184 part 4, triangular pitch 60°, tube 20 × 2, pitch: 25 mm triangular, d_{hole} = 21 mm

Type Nr.	DN	n_P	D_a mm	D_i mm	s mm	SMU mm	D_H mm	B mm	n	n_w	n_{acr}	A m²/m	f_W	VR m³/h	VM m³/h
1	250	2	273	260.4	6.3	1.25	248	50	76	30	10	4.78	0.3983235	27.51	10.27
2	250	4	273	260.4	6.3	1.25	250	50	64	24	8	4.02	0.2880986	11.58	17.30
3	300	4	324	309.8	7.1	1.25	297	60	92	36	12	5.78	0.4417207	16.65	14.48
4	350	8	355	339	8	1.5	328	70	100	28	12	6.28	0.3765783	9.05	24.25
5	400	8	406	388.4	8.8	1.5	373	80	144	32	14	9.05	0.4011764	13.03	30.55
6	500	8	508	496	6	1.75	486	100	256	74	18	16.09	0.3862411	23.16	48.32
7	600	8	600	588	6	1.75	574	120	376	114	22	23.62	0.4213845	34.02	63.31
8	700	8	700	684	8	2.25	671	140	532	176	26	33.43	0.4240278	48.13	82.04
9	800	8	800	784	8	2.25	770	160	724	208	30	45.49	0.4430131	65.51	105.38
10	900	8	900	880	10	2.25	868	180	936	286	34	58.81	0.4594987	84.69	129.00
11	1000	8	1000	980	10	2.75	970	200	1196	390	38	75.15	0.4505112	108.21	157.81
12	1100	8	1100	1076	12	2.75	1060	220	1436	412	42	90.23	0.4751299	129.93	186.32
13	1200	8	1200	1176	12	2.75	1162	240	1736	528	46	109.08	0.4792821	157.07	220.60

Table 3.3 Geometrical data of heat exchangers according to VDI-Wärmeatlas 1984 chap. GDIN 28184 (triangular pitch 60°), tubes 16 × 2, pitch 20 mm triangular, d_{hole} = 17 mm

Type Nr.	DN	n_P	D_a mm	D_i mm	s mm	SMU mm	D_H mm	B mm	n	n_W	n_{acr}	A m²/m	f_W	VR m³/h	VM m³/h
1	150	2	168	159	4.5	1.25	147.4	50	36	10	7	1.81	0.3852249	7.33	7.56
2	200	4	219	207.2	5.9	1.25	198.1	50	60	22	8	3.02	0.2937709	6.11	13.48
3	250	4	273	260.4	6.3	1.25	253.9	50	96	32	10	4.83	0.2797606	9.77	17.47
4	300	8	324	309.8	7.1	1.25	297.7	60	128	42	12	6.43	0.2947632	6.51	24.85
5	350	8	355	339	8	1.5	330.8	70	164	56	14	8.24	0.3090594	8.35	28.39
6	400	8	406	388.4	8.8	1.5	377.8	80	228	68	16	11.46	0.3157251	11.60	7.33
7	500	8	508	496	6	1.75	485	100	404	134	22	20.31	0.3392087	20.56	51.32

Table 3.4 Geometrical data of heat exchanger with flanged floating head according to DIN 28191, tubes 25 × 2, pitch 32 mm triangular, $d_{hole} = 26$ mm (tube hole and the baffle)

Type Nr.	DN	n_P	D_a mm	D_i mm	s mm	SMU mm	D_H mm	B mm	n	n_W	n_{acr}	A m²/m	f_W	VR m³/h	VM m³/h
1	150	2	168.3	159	4.5	1.25	125.6	50	10	2	3	0.8	0.823	6.2	11.1
2	200	2	219.1	207	6.3	1.25	167.7	50	18	3	4	1.4	0.774	11.2	14.5
3	250	2	273	260	6.3	1.25	226.8	50	32	10	5	2.5	0.625	20	16.1
4	300	2	323.9	310	7.1	1.25	268.2	60	46	10	6	3.6	0.656	28.7	24.1
5	350	4	355.6	340	8	1.5	292.8	70	42	5	6	3.3	0.69	13.1	38.8
6	400	4	406.4	389	8.8	1.5	349.3	80	70	20	8	5.5	0.677	21.8	40.9
7	500	4	508	490	8.8	1.75	450	100	122	40	10	9.6	0.709	38	62.8
8	600	8	600	588	6	1.75	539.4	120	172	58	12	13.5	0.723	26.8	96.6
9	700	8	700	684	8	2.25	639	140	248	66	14	19.5	0.71	38.7	127.1
10	800	8	800	784	8	2.25	738.9	160	344	90	16	27	0.709	53.6	169.9
11	900	8	900	880	10	2.25	827.4	180	432	130	18	33.9	0.723	67.3	228.6
12	1000	8	1000	980	10	2.75	931.8	200	560	142	20	44	0.711	87.3	241.3
13	1100	8	1100	1076	12	2.75	1025.2	220	688	174	22	54	0.715	107.2	289.6
14	1200	8	1200	1176	12	2.75	1125.4	240	840	238	24	66	0.704	130.9	351

Example 5: Design of a convective heat exchanger

	Tube side	Shell side
Flow (kg/h)	42,828	28,100
Density (kg/m³)	995	763.4
Specific heat capacity (Wh/kg K)	1.16	0.74
Kinematic viscosity (mm²/s)	0.92	0.9
Conductivity (W/m K)	0.605	0.162
Inlet temperature (°C)	25	58
Outlet temperature (°C)	31.7	42
Fouling factor (m² K/W)	0.0002	0.0002
Heat duty (kW)	332.8	332.8
Prandtl number Pr	6.32	11.3
Volumetric flow (m³/h)	43	36.8

Chosen: Type 10 in Table 1, DN 500 with 164 tubes 25 × 2, triangular pitch 32 mm

Flow rate at 1 m/s	VR = 51.1 m³/h	VM = 42.8 m³/h
Flow velocity	$\frac{43}{51.1}$ = 0.84 m/s	$\frac{36.8}{42.8}$ = 0.86 m/s
Reynolds number Re	19,218	23,472
Nusselt number Nu	112.96	182.85
Heat transfer coefficient α	3254 W/m² K	1185 W/m² K
Overall heat transfer coefficient U	628 W/m² K	
Corrected temperature difference	CMTD = 20.43 °C	

Required heat exchange area A_{req}:

$$A_{req} = \frac{Q}{U \times CMTD} = \frac{332,800}{628 \times 20.43} = 25.9 \text{ m}^2$$

Chosen: tube length L = 2.5 m
Chosen heat exchange area A = 2.5 m × 12.88 m²/m = 33.2 m²
Reserve: 24%

Example 6: Design of a heat exchanger

	Tube side	Shell side
Flow (kg/h)	14,276	28,100
Density (kg/m³)	995	763.4
Specific heat capacity (Wh/kg K)	1.16	0.74
Kinematic viscosity (mm²/s)	0.92	0.9
Conductivity (W/m K)	0.605	0.162
Inlet temperature (°C)	15	58
Outlet temperature (°C)	35.1	42
Fouling factor (m² K/W)	0.0002	0.0002

Heat load (kW)	332.8	332.8
Prandtl number Pr	6.32	11.3
Volumetric flow (m³/h)	14.35	36.8

LMTD = 24.89 °C
CMTD = 22.56 °C

LMTD, logarithmic mean temperature difference; CMTD, corrected effective mean temperature difference.

Chosen: Type 10 in Table 1: DN 500 with 164 tubes 25 × 2, triangular pitch 32 mm

Flow rate at 1 m/s	$VR = 51.1 \text{ m}^3/\text{h}$	$VM = 42.8 \text{ m}^3/\text{h}$
Flow velocity	$\frac{14.35}{51} = 0.28 \text{ m/s}$	$\frac{36.8}{42.8} = 0.86 \text{ m/s}$

The flow velocity on the tube side is too small.

New choice: Type 4 in Table 2: DN 350 with 100 tubes 20 × 2, triangular pitch 25 mm

Flow rate at 1 m/s	$VR = 9 \text{ m}^3/\text{h}$	$VM = 24.2 \text{ m}^3/\text{h}$
Flow velocity	$\frac{14.35}{9} = 1.56 \text{ m/s}$	$\frac{36.8}{24.2} = 1.53 \text{ m/s}$
Reynolds number Re	27,130	34,074
Nusselt number Nu	148.8	228.7
Heat transfer coefficient α (W/m² K)	5600	1852
Overall heat transfer coefficient	$U = 863 \text{ W/m}^2 \text{ K}$	

Required heat exchange area A_{req}:

$$A_{req} = \frac{332,800}{863 \times 22.56} = 17 \text{ m}^2$$

Chosen tube length $L = 3.5$ m
Chosen heat exchange area $A = 3.5 \times 6.28 = 22 \text{ m}^2$
29 % reserve.

Notes for the heat exchanger tables:

A = heat exchanger surface area per m tube length (m²/m)
B = baffle spacing (mm)
D_a = outer diameter (mm)
D_i = inner diameter (mm)
D_H = bundle outer limit (mm)
d_{hole} = hole diameter in baffle plate (mm)
f_W = correction factor according to VDI-Wärmeatlas
n = number of tubes
n_{acr} = number of tubes in cross stream
n_P = number of tube passes
n_w = number of tubes in baffle window
s = shell wall thickness (mm)
SMU = width of gap between baffle and shell inner diameter (mm)
VR = required volumetric flow (m³/h) for 1 m/s flow velocity on the tube side
VM = required volumetric flow (m³/h) for 1 m/s flow velocity on the shell side

NOMENCLATURE

a_{cross} flow cross section shell side (m^2)

a_T flow cross section tube side (m^2)

B baffle spacing (m)

D_i shell inner diameter (m)

d_i tube inner diameter (m)

d_o tube outer diameter (m)

l tube length (m)

n_{acr} number of tubes in cross stream

n_P number of tube passes

w flow velocity (m/s)

w_{cross} cross flow velocity shell side (m/s)

w_T tube-side velocity (m/s)

V volumetric throughput (m^3/h)

Nu Nusselt number

Re Reynolds number

Pr Prandtl number

T pitch (m)

α heat transfer coefficient $(W/m^2 K)$

c specific heat capacity (Wh/kg K)

λ heat conductivity (W/m K)

ν kinematic viscosity (m^2/s)

ρ density (kg/m^3)

REFERENCES AND FURTHER READING

[1] R. Mukherjes, Chem. Eng. Prog. 92 (1996) 72–79 + 94 (1998) 21–37.
[2] J. Fisher, R.O. Parker, Hydrocarbon Process. 48 (1969) 147–154.
[3] A.P. Fraas, M.N. Ozisik, Heat Exchanger Design, John Wiley, N.Y, 1965.
[4] A.E. Jones, Thermal design of the shell-and-tube, Chem. Eng. 109 (2002) 60–65.
[5] A. Devore, G.J. Vago, G.J. Picozzi, Specifying and selecting, Chem. Eng. (Octomber 1980).
[6] E.N. Sieder, G.E. Tate, Ind. Eng. Chem. 28 (1936) 1429.
[7] D.A. Donohue, Heat transfer and pressure drop in heat exchangers, Ind. Eng. Chem. 41 (1949) 2499.
[8] Petr.Ref. 34 (1955), No. 10 und No. 11.
[9] T. Tinker, Shell-side characteristics of shell and tube heat exchangers, Trans. ASME 80 (1958) 36 (ff).
[10] K.J. Bell, Thermal design of heat transfer equipment, in: Perry's Handbook, 1973.
[11] VDI-Wärmeatlas, Aufkage, 5, VDI-Verlag, Düsseldorf, 1988.
[12] Lord, Minton, Slusser, Design of heat exchangers, Chem. Eng, 26 (January 1970).
[13] M. Jakob, Trans. ASME 60 (1938), 381/392.
[14] Coates, B.S. Pressburg, Heat transfer to moving fluids, Chem. Eng. Dez. 66 (1959) 67–72.
[15] N.P. Chopey, Heat transfer, in: Handbook of Chemical Engineering Calculations, McGraw-Hill, New York, 1993.
[16] D.Q. Kern, Process Heat Transfer, McGraw-Hill, N.Y. 1950.
[17] L. Clarke, R.L. Davidson, Manual for Process Engineering Calculations, McGraw-Hill, N.Y, 1962.

CHAPTER 4

Geometrical Heat Exchanger Calculations

Contents

4.1 CALCULATION FORMULA

4.1.1 Heat exchanger area A

$$A = n \times d_o \times \pi \times L \, (\mathrm{m}^2)$$

n = total number of tubes
d_o = tube outer diameter (m)
L = tube length (m)

© 2016 Elsevier Inc.
All rights reserved.

4.1.2 Tube-side flow cross section a_T

$$a_T = \frac{n}{n_P} \times d_i^2 \times \frac{\pi}{4} \; (m^2)$$

t_P = tubes per pass
d_i = tube inner diameter (m)

4.1.3 Flow velocity w_T in the tubes

$$w_T = \frac{V_T}{a_T \times 3600} \; (m/s)$$

V_T = tube-side volumetrical flow (m^3/h)

4.1.4 Shell-side cross section a_{cross}

$$a_{cross} = B \times (D_i - n_{acr} \times d_o) \; (m^2) \qquad n_{acr} = \frac{D_i - s_R}{s_q} = \frac{D_H}{s_q}$$

B = baffle spacing, ca. $0.2 \times D_i$ (m)
D_i = shell inner diameter (m)
n_{acr} = number of tubes in the middle across the flow direction
s_R = edge strip (m)
s_q = tube middle gap across the flow direction (m) = pitch T
D_H = outer limit diameter (m) = envelope of tube pattern

4.1.5 Shell-side cross stream velocity w_C

$$w_{cross} = \frac{V_{Shell}}{a_{cross} \times 3600} \; (m/s)$$

V_{Shell} = shell-side volumetric flow (m^3/h)

4.1.6 Shell-side longitudinal stream cross area a_w in the baffle window

$$a_w = D_i^2 \times C - n_w \times d_o^2 \times \frac{\pi}{4} \; (m^2) \qquad C = \frac{16 \times x^2 - 13 \times x^3}{12 \times \sqrt{x - x^2}} \qquad \text{with } x = \frac{H}{D_i}$$

C = geometrical calculation factor
H = segment height (m)
n_w = number of tubes in the baffle window
The number of tubes n_w lying in the baffle cut can be determined by calculation from the segment height H and the distance of the tube rows from the middle.

A tube layout is more appropriate.

According to the equation above, the tubes lying in the segment have to be subtracted from the segment area.

As a first approximation, the window area which is not covered with the tubes can be estimated with the free room factor NF.

$$NF = 1 - \frac{n \times d_o^2}{D_i^2} \qquad a_w = D_i^2 \times C \times NF \, (m^2)$$

n = total number of tubes

The calculation with NF gives a little lower flow cross sections in the window because the edge strip was not considered.

In order to reduce the effect of the edge strip, especially with small shell diameters, the outer limit diameter D_H should be used instead of the shell inner diameter D_i.

$$NF = 1 - \frac{n \times d_o^2}{D_H^2} \qquad a_w = D_i^2 \times C \times NF \, (m^2)$$

For the shell side, the following designs should be strived for:

$$a_{cross} = a_w \qquad x = H/D_i = 0.2 \text{ bis } 0.25 \qquad B/D_i = 0.2$$

Calculation factors C for different x-values:

$x = H/D_i$	C
0.20	0.1117
0.21	0.1197
0.22	0.1279
0.23	0.1363
0.24	0.1448
0.25	0.1534
0.26	0.1621
0.27	0.1709
0.28	0.1798
0.29	0.1889
0.30	0.1980

Longitudinal stream flow velocity w_w in the window $\qquad w_w = \dfrac{V_{Shell}}{a_w \times 3600} \, (m/s)$

Average shell-side flow cross section a_{av} $\qquad a_{av} = \sqrt{a_{cross} \times a_w} \, (m^2)$

Average shell-side flow velocity w_{av} $\qquad w_{av} = \dfrac{V_{Shell}}{a_{av} \times 3600} \, (m/s)$

Number of baffles n_B $\qquad n_B = \dfrac{L - B}{B + a}$

a = baffle thickness (m)
B = baffle spacing (m)
L = tube length (m)

$$\text{Check calculation for the baffle spacing } B \quad B = \frac{L - n_B \times a}{n_B + 1} \, (\text{m})$$

4.1.7 Number of tube rows in cross stream n_{cross} between the windows

$$n_{cross} = \frac{D_i - 2H - S_G \times (n_P - 1)}{s_l} + 1$$

n_P = number of tube passes
S_G = strips of the pass partition plates for the pass partition
s_l = tube middle clearance longitudinal to flow direction
Cross and longitudinal clearances $(T = \text{tube pitch})$

Tube middle spacing	Triangular	Square	Rotated square 45°
Cross flow s_q	$s_q = T$	$s_q = T$	$s_q = 1.414\,T$
Longitudinal flow s_l	$s_l = 0.866\,T$	$s_l = T$	$s_l = 0.707\,T$

Triangular   Square   Rotated square

4.2 TUBE-SIDE CALCULATIONS (FIGURE 4.1)

4.2.1 Flow velocity w_T in the tubes for a volumetric flow V_T (m³/h)

$$w_T = \frac{V_T}{2827 \times d_i^2 \times t_P} \, (\text{m/s}) \qquad V_T = \frac{M_T}{\rho_T} \, (\text{m}^3/\text{h}) \qquad t_P = \frac{n}{n_P}$$

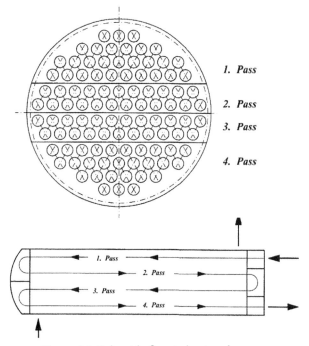

Figure 4.1 Tube-side flow in heat exchanger.

4.2.2 Number of tubes per pass t_P for a given flow velocity w_T

$$t_P = \frac{V_T}{2827 \times d_i^2 \times w_T}$$

4.2.3 Required number of tubes per pass t_P for a given Reynolds number Re

$$t_P = \frac{V_T}{2827 \times \nu \times \mathrm{Re} \times d_i}$$

4.2.4 Required flow velocity w_T for a given Reynolds number Re

$$w_T = \frac{\mathrm{Re} \times \nu}{d_i}\,(\mathrm{m/s})$$

d_i = tube inner diameter (m)
ρ = density $(\mathrm{kg/m^3})$
ν = kinematic viscosity $(\mathrm{m^2/s})$
V_T = volumetric throughput $(\mathrm{m^3/h})$
M_T = mass throughput $(\mathrm{kg/h})$
Re = Reynolds number

n_P = number of tube passes

t_P = Tubes per pass

Example 1: Tube-side calculations

$$M_T = 5000 \text{ kg/h} \qquad \rho = 850 \text{ kg/m}^3 \qquad V_T = 5000/850 = 5.88 \text{ m}^3/\text{h}$$
$$\nu = 2 \text{ mm}^2/\text{s} \qquad d_i = 16 \text{ mm}$$

Calculation of the required flow velocity for Re = 5000

$$w_T = \frac{\text{Re} \times \nu}{d_i} = \frac{5000 \times 2 \times 10^{-6}}{0.016} = 0.625 \text{ m/s}$$

Required number of tubes t_P per pass for w_T = 0.625 m/s

$$t_P = \frac{5.88}{2827 \times 0.016^2 \times 0.625} = 13 \text{ Rohre mit } d_i = 16 \text{ mm}$$

Required number of tubes t_P per pass for Re = 5000

$$t_P = \frac{5.88}{2827 \times 2 \times 10^{-6} \times 5000 \times 0.016} = 13 \text{ Rohre}$$

4.3 SHELL-SIDE CALCULATIONS (FIGURE 4.2)

4.3.1 Flow cross section a_{cross} for the flow across the tubes (Figure 4.2)

$$a_{cross} = B \times (D_i - n_{acr} \times d_o)(\text{m}^2)$$

B = baffle spacing (m)

d_o = outer diameter of the tubes (m)

D_i = shell inner diameter (m)

D_H = outer limit diameter (m)

n_{acr} = number of tubes in cross flow on the center line

T = pitch (m)

$s = D_i - D_H$ = edge strips between shell and tube bundle

4.3.2 Flow cross section a_w for the longitudinal flow in the baffle cuts

$$a_w = A_S - n_W \times d_o^2 \times \pi/4$$

A_S = area of the baffle cut (m^2)

$$\text{For } H/D_i = 0.2 \rightarrow A_S = 0.112 \times D_i^2$$
$$\text{For } H/D_i = 0.25 \rightarrow A_S = 0.154 \times D_i^2$$

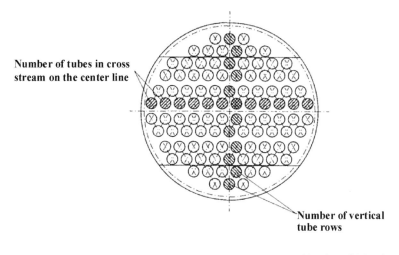

Number of tubes in cross
stream on the center line

Number of vertical
tube rows

Number of tubes in
the baffle window

Segment height H = 80 mm

Segment height H = 80 mm

Number of tube rows
between windows

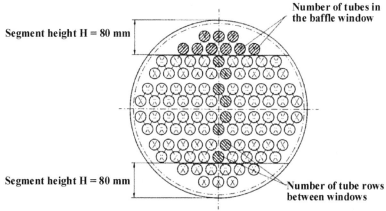

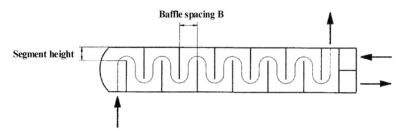

Baffle spacing B

Segment height

Figure 4.2 Shell-side flow in heat exchanger.

n_w = number of tubes in the baffle window

H = segmental height = height of the baffle cut

If the number of tubes in the segment is not known, the longitudinal stream cross section a_w can be determined with the NF factor.

$$\mathrm{NF} = 1 - \frac{n \times d_o^2}{D_i^2} \quad n = \text{total number of tubes}$$

$$a_W = \mathrm{NF} \times A_S \,(\mathrm{m}^2)$$

Notice: Equal flow cross sections will be preferably desired for the cross and longitudinal flow.

4.3.3 Average flow cross section a_{av} on the shell side

$$a_{av} = \sqrt{a_{cross} \times a_W} \,(\mathrm{m}^2)$$

4.3.4 Shell-side flow velocities w_{av}

$$\text{Average velocity } w_{av} = \frac{V_{Shell}}{3600 \times a_{av}} = \frac{V_{Shell}}{3600 \times \sqrt{a_{across} \times a_w}} \,(\mathrm{m/s})$$

$$w_{cross} = \frac{V_{Shell}}{3600 \times a_{cross}} \,(\mathrm{m/s}) = \text{crossflow velocity through the bundle}$$

$$w_w = \frac{V_{Shell}}{3600 \times a_w} \,(\mathrm{m/s}) = \text{longitudinal flow velocity through the window}$$

4.3.5 Required baffle spacing B for a given cross stream velocity w_{cross}

$$B = \frac{V_{Shell}}{3600 \times w_{cross} \times (D_i - n_{acr} \times d_o)} \,(\mathrm{m})$$

4.3.6 Number of baffles n_B with given baffle spacing B

$$n_B = \frac{L - B}{B + a} \qquad B = \frac{L - n_B \times a}{n_B + 1} \,(\mathrm{m})$$

4.3.7 Determination of the total vertical tube rows n_{tv}

$$n_{tv} = \frac{D_i - d_o - S_G \times (n_p - 1)}{s_1}$$

S_G = stripes for pass partitions

n_P = number of tube passes

4.3.8 Determination of the tube rows n_{cross} in the cross stream between the baffle windows with the segment height H

$$n_{\text{cross}} = \frac{D_{\text{H}} - d_{\text{o}} - 2H - (n_{\text{P}} - 1) \times S_{\text{G}}}{s_1} + 1$$

Example 2: Shell-side Calculations

Heat exchanger DN 400 with 106 tubes 25×2 in triangular pitch $T = 32$ mm

$D_{\text{i}} = 400$ mm	$L = 4$ m	$a = 5$ mm	$n_{\text{P}} = 4$	$S_{\text{G}} = 25$ mm
$D_{\text{H}} = 373$ mm for	$n_{\text{P}} = 2$	$n = 106$	$A = 8.32$ m^2/m	
$D_{\text{H}} = 369$ mm for	$n_{\text{P}} = 4$	$n = 88$	$A = 6.91$ m^2/m	

Baffle spacing $B = 0.2 \times D_{\text{i}} = 0.2 \times 400 = 80$ mm

Segment height $H = 0.2 \times D_{\text{i}} = 0.2 \times 400 = 80$ mm

$H/D_{\text{i}} = 0.2 \Rightarrow A_{\text{S}} = 0.112 \times D_{\text{i}}^2 = 0.112 \times 0.4^2 = 0.0179$ m^2

Calculation of the flow sections:

$$n_{\text{acr}} = \frac{D_{\text{i}}}{T} = \frac{400}{32} = 12 \text{ Tubes} \qquad NF = 1 - \frac{n \times d_{\text{o}}^2}{D_{\text{i}}^2} = 1 - \frac{106 \times 0.025^2}{0.4^2} = 0.586$$

$$a_{\text{cross}} = B \times (D_{\text{i}} - n_{\text{acr}} \times d_{\text{o}}) = 0.08 \times (0.4 - 12 \times 0.025) = 0.008 \text{ m}^2$$

$$a_{\text{w}} = NF \times A_{\text{S}} = 0.586 \times 0.0179 = 0.0105$$

$$a_{\text{av}} = \sqrt{0.008 \times 0.0105} = 0.0092 \text{ m}^2$$

Calculation of the shell-side flow velocities for $V_{\text{Shell}} = 20$ m^3/h

$$w_{\text{cross}} = \frac{20}{3600 \times 0.008} = 0.694 \text{ m/s} \qquad w_{\text{w}} = \frac{20}{3600 \times 0.0105} = 0.53 \text{ m/s}$$

$$w_{\text{av}} = \frac{20}{3600 \times 0.0092} = 0.604 \text{ m/s}$$

Determination of the number of baffles n_{B} for $a = 5$ mm

$$n_{\text{B}} = \frac{L - B}{B + a} = \frac{4000 - 80}{80 + 5} = 46 \text{ baffles}$$

Check calculation of the baffle spacing B for $a = 5$ mm

$$B = \frac{L - n_{\text{B}} \times a}{n_{\text{B}} + 1} = \frac{4000 - 46 \times 5}{46 + 1} = 80 \text{ mm}$$

Estimation of the total vertical tube rows n_{tv} for $n_{\text{P}} = 2$

$$n_{\text{TV}} = \frac{D_{\text{H}} - d_{\text{o}} - S_{\text{G}} \times (n_{\text{P}} - 1)}{0.866 \times T} = \frac{373 - 25 - 10 \times (2 - 1)}{0.866 \times 32} = 12.2$$

Only even numbers are possible: n_{tv} = vertical 12 tube rows.

Determination of the vertical tube rows n_{cross} between the baffle cuts for $n_P = 2$

$$n_{cross} = \frac{D_H - d_o - (n_P - 1) \times S_G - 2 \times H}{0.866 \times T} + 1$$

$$= \frac{373 - 25 - (2 - 1) \times 10 - 2 \times 80}{0.866 \times 32} + 1 = 7.42$$

Only even numbers are possible: n_{cross} = vertical seven tube rows between the windows.

Example 3: Calculation of n_{cross} for the data in Figure 4.3

$$n_{cross} = \frac{442 - 25 - (4 - 1) \times 15 - 2 \times 109.6}{22.6} + 1 = 7.76$$

Only even numbers are possible: n_{cross} = vertical seven tube rows between the windows.

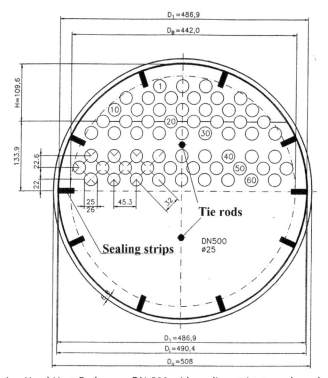

Figure 4.3 Floating Head Heat Exchanger DN 500 with sealing strips to reduce the bypass stream.

CHAPTER 5

Dimensionless Characterization Number for the Heat Transfer

Contents

5.1 REYNOLDS NUMBER RE FOR THE CHARACTERIZATION OF THE FLOW CONDITION

$$\text{Re} = \frac{w \times d}{\nu}$$

w = flow velocity (m/s)
d = tube diameter (m)
ν = kinematic viscosity

Example 1: Calculation of the Reynolds number Re

Data: $w = 1.2$ m/s $\quad d_i = 21$ mm $\quad \nu = 0.95 \times 10^{-6}$ m²/s

$$\text{Re} = \frac{1.2 \times 0.021}{0.95 \times 10^{-6}} = 26,526$$

5.2 PRANDTL NUMBER PR, PECLET NUMBER PE, AND TEMPERATURE CONDUCTIVITY A [1−8]

$$\text{Pr} = \frac{\nu}{a} = \frac{\text{Pe}}{\text{Re}} = \frac{3600 \times \eta \times c_p}{\lambda} = \frac{3600 \times \nu \times c_p \times \rho}{\lambda}$$

$$\text{Pe} = \frac{w \times d}{a} = \text{Pr} \times \text{Re} \quad a = \frac{\lambda}{c_p \times \rho \times 3600} \ (\text{m}^2/\text{s})$$

Heat Exchanger Design Guide
http://dx.doi.org/10.1016/B978-0-12-803764-5.00005-5

© 2016 Elsevier Inc.
All rights reserved.

a = temperature conductivity (m²/s)
Pe = Peclet number (−)
ρ = density (kg/m³)
c_p = specific heat capacity (Wh/kg K)
λ = thermal heat conductivity (W/m K)
η = dynamic viscosity (Pas)
ν = kinematic viscosity (m²/s)

The factor 3600 is needed to convert hours to seconds in order to make the dimensions coherent.

Example 2: Calculation of the Prandtl number Pr

Data: $\rho = 903$ kg/m³ $c_p = 0.594$ Wh/kg K $\lambda = 0.118$ W/m K
  $\eta = 858$ mPas $\nu = 0.95 \times 10^{-6}$ m²/s

$$a = \frac{\lambda}{c_p \times \rho \times 3600} = \frac{0.118}{0.594 \times 903 \times 3600} = 6.1 \times 10^{-8} \text{ m}^2/\text{s}$$

$$Pr = \frac{\nu}{a} = \frac{0.95 \times 10^{-6}}{6.1 \times 10^{-8}} = 15.5$$

$$Pr = \frac{3600 \times \nu \times c_p \times \rho}{\lambda} = \frac{3600 \times 0.95 \times 10^{-6} \times 0.594 \times 903}{0.118} = 15.5$$

$$Pr = \frac{3600 \times \eta \times c_p}{\lambda} = \frac{3600 \times 858 \times 10^{-6} \times 0.594}{0.118} = 15.5$$

5.3 NUSSELT NUMBER NU FOR THE CALCULATION OF THE HEAT TRANSFER COEFFICIENT [6]

$$Nu = \frac{\alpha \times d}{\lambda} = St \times Re \times Pr \quad \alpha = \frac{Nu \times \lambda}{d} \left(W/m^2 K \right)$$

α = heat transfer coefficient (W/m² K)

5.4 STANTON NUMBER ST FOR THE CALCULATION OF THE HEAT TRANSFER COEFFICIENT

$$G = w \times \rho \text{ (kg/m}^2 \text{ s)} \qquad G = \text{mass stream density (kg/m}^2 \text{ s)}$$

$$St = \frac{Nu}{Re \times Pr} = \frac{\alpha}{G \times c} = \frac{\alpha}{3600 \times c \times w}$$

$$a = St \times 3600 \times c_p \times w \times \rho \ (m^2/s)$$

5.5 COLBURN FACTOR J_C FOR THE CALCULATION OF THE HEAT TRANSFER COEFFICIENT

$$J_C = St \times Pr^{2/3} \quad \alpha = \frac{J_C}{Pr^{2/3}} \times 3600 \times w \times \rho \ (W/m^2 \ K)$$

5.6 KERN FACTOR J_K FOR THE CALCULATION OF THE HEAT TRANSFER COEFFICIENT [1]

$$J_K = \frac{Nu}{Pr^{1/3}} \quad \alpha = J_K \times Pr^{1/3} \times \frac{\lambda}{d} \ (W/m^2 \ K)$$

Example 3: Calculation of the heat transfer coefficient with Nu, St, J_C, and J_K

Data: $\rho = 903 \ kg/m^3$ $c_p = 0.594 \ Wh/kg \ K$ $\lambda = 0.118 \ W/m \ K$
$\nu = 0.95 \times 10^{-6} \ m^2/s$ $d_i = 21 \ mm$ $w = 1.2 \ m/s$
Re $= 26526$ Pr $= 15.5$ Nu $= 196.55$

$$\alpha = \frac{Nu \times \lambda}{d} = \frac{196.55 \times 0.118}{0.021} = 1104 \ W/m^2 \ K$$

$$St = \frac{Nu}{Re \times Pr} = \frac{196.55}{26526 \times 15.5} = 0.00048$$

$$\alpha = St \times 3600 \times w \times \rho = 0.00048 \times 3600 \times 0.594 \times 1.2 = 1108 \ W/m^2 \ K$$

$$J_C = St \times Pr^{2/3} = 0.00048 \times 15.5^{2/3} = 0.003$$

$$\alpha = \frac{J_C}{Pr^{2/3}} \times 3600 \times c_p \times w \times \rho = \frac{0.003}{15.5^{2/3}} \times 3600 \times 0.594 \times 1.2 \times 903$$
$$= 1112 \ W/m^2 \ K$$

$$J_K = \frac{Nu}{Pr^{1/3}} = \frac{196.55}{15.5^{1/3}} = 79.55$$

$$\alpha = J_K \times Pr^{1/3} \times \frac{\lambda}{d} = 79.55 \times 15.5^{1/3} \times \frac{0.118}{0.021} = 1104 \ W/m^2 \ K$$

5.7 GRAßHOF NUMBER GR FOR THE CALCULATION OF THE HEAT TRANSFER COEFFICIENT IN NATURAL CONVECTION [8]

$$\mathrm{Gr} = \frac{9.81 \times \beta \times \Delta t \times l^3}{\nu} \quad \beta = \frac{1}{\nu} \times \frac{\nu_1 - \nu_2}{t_1 - t_2} = \rho \times \frac{\frac{1}{\rho_1} - \frac{1}{\rho_2}}{t_1 - t_2} \ (1/K)$$

$\beta =$ volumetric heat expansion coefficient (1/K)
$\Delta t =$ temperature difference product—wall (K)
$l =$ characteristic length (m)
$\nu_1 =$ specific volume (m^3/kg) at t_1
$\nu_2 =$ specific volume (m^3/kg) at t_2
Product Gr·Pr

$$\mathrm{Gr} \times \mathrm{Pr} = 35316 \times \frac{\rho \times c_{\mathrm{p}} \times \beta \times l^3 \times \Delta t}{\lambda \times \nu}$$

Example 4: Calculation of the Graßhof number

Data: $\rho = 800$ kg/m^3 $c_{\mathrm{p}} = 0.6$ Wh/kg K $\lambda = 0.13$ W/m K

$\nu = 50 \times 10^{-6}$ m^2/s $\beta = 0.001$ K^{-1} $\Delta t = 50$ K $d = 50$ mm

$$\mathrm{Pr} = \frac{3600 \times 50 \times 10^{-6} \times 0.6 \times 800}{0.13} = 664.6$$

$$\mathrm{Gr} = \frac{9.81 \times 0.001 \times 50 \times 0.05^3}{50 \times 10^{-6}} = 24.53 \times 10^3$$

$$\mathrm{Gr} \times \mathrm{Pr} = 24.53 \times 10^3 \times 664.6 = 16.3 \times 10^6$$

$$\mathrm{Gr} \times \mathrm{Pr} = 35316 \times \frac{800 \times 0.6 \times 0.002 \times 0.05^3 \times 50}{50 \times 10^{-6} \times 0.13} = 16.3 \times 10^6$$

REFERENCES

[1] D.Q. Kern, Process Heat Transfer, McGraw-Hill, N.Y., 1950.
[2] A.P. Fraas, M.N. Ozisik, Heat Exchanger Design, John Wiley, N.Y., 1965.
[3] W.H. McAdams, Heat Transmission, McGraw-Hill, N.Y., 1954.
[4] Perry's Chemical Engineer's Handbook, McGraw-Hill, 1984.
[5] G.F. Hewitt, G.L. Shires, T.R. Bott, Process Heat Transfer, CRC Press, Boca Raton, 1994.
[6] V.D.I. Wärmeatlas, 5. Auflage, VDI-Verlag, Düsseldorf, 1988.
[7] W.M. Rohsenow, J.P. Hartnett, Handbook of Heat Transfer, McGraw-Hill, N.Y., 1973.
[8] E.R.G. Eckert, R.M. Drake, Heat and Mass Transfer, McGraw Hill, N.Y., 1959.

CHAPTER 6

Overall Heat Transfer Coefficient and Temperature Profile

Contents

6.1 CALCULATION OF THE OVERALL HEAT TRANSFER COEFFICIENT [1–10]

The equation for the heat transfer is as follows:

$$Q = U \times A \times \text{LMTD} \quad A = \frac{Q}{U \times \text{LMTD}}$$

Normally the required heat exchanger surface area A is calculated for a given problem definition. The amount of heat Q to be transferred will result from the problem definition.

Likewise, the logarithmic mean temperature difference, LMTD, or for nonideal countercurrent flow the corrected mean effective temperature difference (CMTD).

The main problem lies in establishing the heat transfer coefficients on the tube and shell side of the heat exchanger.

If the heat transfer coefficients are known, the overall heat transfer coefficient can be established.

Without considering the different areas on the inside and outside of the heat exchanger tubes and the fouling and the conduction through the tube wall the following U_α-value follows from the two heat transfer coefficients:

$$U_\alpha = \frac{1}{\frac{1}{\alpha_i} + \frac{1}{\alpha_o}} = \frac{\alpha_i \times \alpha_o}{\alpha_i + \alpha_o}$$

Heat Exchanger Design Guide
http://dx.doi.org/10.1016/B978-0-12-803764-5.00006-7
© 2016 Elsevier Inc.
All rights reserved.

Example 1: Calculation of U_α from the heat transfer coefficients

Data: $\alpha_i = 620$ W/m^2 K $\alpha_o = 699$ W/m^2 K

$$U_\alpha = \frac{620 \times 699}{620 + 699} = 328 \text{ W/m}^2 \text{ K}$$

Alternatively the overall heat transfer coefficient U_α can be determined from the diagram in Figure 6.1.

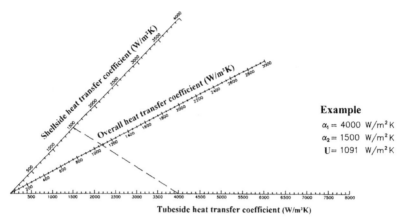

Figure 6.1 Diagram for the determination of U_α from the α-values.

Example 2:

Data: $\alpha_i = 4000$ W/m^2 K $\alpha_o = 1500$ W/m^2 K

$$U_\alpha = \frac{4000 \times 1500}{4000 + 1500} = 1091 \text{ W/m}^2 \text{ K}$$

The good heat transfer coefficient in the tube does not help much. The overall heat transfer coefficient is always determined by the worse heat transfer coefficient.

The inner heat transfer coefficient α_i must be converted to α_{io} because with plain tubes the heat exchanger surface area generally refers to the outer area of the tube.

$$\alpha_{io} = \alpha_i \times \frac{d_i}{d_o} = \alpha_i \times \frac{A_i}{A_o} \ (\text{W/m}^2 \text{ K})$$

Example 3: Conversion of α_i to the outside area

Data: $\alpha_i = 620$ W/m^2 K $d_i = 16$ mm $d_a = 20$ mm

$$\alpha_{io} = 620 \times \frac{16}{20} = 496 \text{ W/m}^2\text{K}$$

The heat transfer coefficient α_{io} is the tube side heat transfer coefficient based on the outside area of the tubes.

Next the heat conduction through the tube wall is to be considered especially with tube materials having bad heat conductivity λ, for instance, stainless steel.

$$U_{clean} = \frac{1}{\frac{1}{\alpha_{io}} + \frac{1}{\alpha_o} + \frac{s}{\lambda}} \quad (W/m^2\,K)$$

Example 4: Calculation of U_{clean}

Data: $\alpha_{io} = 496\ W/m^2\,K$ $\alpha_o = 699\ W/m^2\,K$ $s = 2\ mm$ $\lambda = 56\ W/m\,K$

$$U_{clean} = \frac{1}{\frac{1}{496} + \frac{1}{699} + \frac{0.002}{56}} = 287\ W/m^2\,K$$

If chromium–nickel–steel is used with a heat conductivity of 16 W/m K, this deteriorates the overall heat transfer coefficient.

$$U_{clean} = \frac{1}{\frac{1}{496} + \frac{1}{699} + \frac{0.002}{16}} = 280\ W/m^2\,K$$

Finally, the fouling f_i on the tube inside and f_a on the tube outside have to be considered.

$$U_{dirty} = \frac{1}{\frac{1}{\alpha_{io}} + \frac{1}{\alpha_o} + \frac{s}{\lambda} + f_i + f_a} = \frac{1}{\frac{1}{U_{clean}} + f_i + f_a} \quad (W/m^2\,K)$$

The fouling is dependent on the products, the operation period, the flow velocity, and the performed cleanings. The values for the fouling f_i on the tube inside and f_a on the shell side lie in the range of 0.0001 −0.001 m² K/W depending on the problem definition.

Material	<1 m/s		>1 m/s	
	<50 °C	>50 °C	<50 °C	>50 °C
Drinking water	0.0002	0.0004	0.0002	0.0004
Distilled water	0.0001	0.0001	0.0001	0.0001
River water	0.0006	0.0008	0.0004	0.0006
Cooling water	0.0002	0.0004	0.0002	0.0004
Brackish water	0.0004	0.0006	0.0002	0.0004
Sea water	0.0001	0.0002	0.0001	0.0002
Vessel-desalted water	0.0004	0.0004	0.0004	0.0004

Material	f (m² K/W)
Fuel oil	0.001
Lubricating oil	0.0002
Thermal oil	0.0002
Vegetable oil	0.0006
Organic liquids	0.0002
Organic vapors	0.0001
Diesel-exhaust gas	0.002
Oil-free vapor	0.0001
Oil-rich vapor	0.0002
Air	0.0004

The values are valid for an operation period of 1 year.

If the tubes will be coated for corrosion reasons, the heat conductivity resistance of the coating must be additionally considered.

Example 5: Calculation of U_{dirty}

Data: $\alpha_{io} = 496$ W/m^2 K $\alpha_o = 699$ W/m^2 K $s = 2$ mm $\lambda = 56$ W/m K
$f_i = 0.0002$ m^2 K/W $f_a = 0.0002$ m^2 K/W $U_{clean} = 287$ W/m^2 K

$$U_{dirty} = \frac{1}{\frac{1}{496} + \frac{1}{699} + \frac{0.002}{56} + 0.0002 + 0.0002} = 257 \text{ W/m}^2 \text{ K}$$

$$U_{dirty} = \frac{1}{\frac{1}{287} + 0.0002 + 0.0002} = 257 \text{ W/m}^2 \text{ K}$$

It might be interesting to establish the fouling reserve of a new heat exchanger.

Thereby, the determined U_{clean}-value from the two heat transfer coefficients and the heat conduction through the tubes is compared with that U_{req}-value required for the problem definition.

$$\sum f = \frac{U_{clean} - U_{eq}}{U_{clean} \times U_{req}}$$

Example 6: Calculation of the existing fouling reserve Σf

Data: $\alpha_{io} = 496$ W/m^2 K $\alpha_o = 699$ W/m^2 K $U_{clean} = 287$ W/m^2 K
$s = 2$ mm $\lambda = 56$ W/m K $U_{req} = 230$ W/m^2 K

$$\sum f = \frac{U_{clean} - U_{eq}}{U_{clean} \times U_{req}} = \frac{287 - 230}{287 \times 230} = 0.0009 \frac{\text{m}^2 \text{ K}}{\text{W}}$$

In this case, for both fouling values $f_i + f_a = 0.0004$ m^2 K/W is used so that an additional fouling reserve of $0.0009 - 0.0004 = 0.0005$ m^2 K/W is available.

Influences of fouling

In Figure 6.2 it is shown that a good overall heat transfer coefficient U_{clean} is more highly reduced through fouling than a bad overall heat transfer coefficient U_{clean}.

Through the fouling the U-values are leveled.

The calculations so far are based on the assumption of plain walls.

If the tube form in shell and tube heat exchangers is considered, then the following exact equation for the overall heat transfer coefficient U_{ex} in tubes results:

$$U_{ex} = \frac{1}{\left(f_i \times \frac{1}{\alpha_i} \right) \times \frac{d_o}{d_i} + \frac{d_o}{2 \times \lambda} \times \ln \frac{d_o}{d_i} + \frac{1}{\alpha_o} + f_a}$$

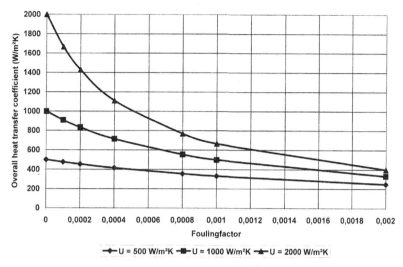

Figure 6.2 The fall of the overall heat transfer coefficient with increasing fouling.

Example 7: Calculation of U_{ex} for the data of Example 5

$$U_{ex} = \cfrac{1}{\left(0.0002 \times \frac{1}{620}\right) \times \frac{20}{16} + \frac{0.02}{2\times56} \times \ln\frac{20}{16} + \frac{1}{699} + 0.0002} = 254 \text{ W/m}^2 \text{ K}$$

The overall heat transfer coefficient U_{ex} for tubes is a little bit lower, but the difference between U_{ex} and U_{dirty} is small, so that a simple equation for plain walls can be used for the calculation.

Recommended calculation procedure
1. Calculation of the heat transfer coefficients α_i in the tube and α_o on the shell side
2. Correction of α_i to α_{io}: $\alpha_{io} = \alpha_i \times \frac{d_i}{d_o}$ (W/m² K)
3. Calculation of U_{clean}: $U_{clean} = \cfrac{1}{\frac{1}{\alpha_{io}} + \frac{1}{\alpha_o} + \frac{s}{\lambda}}$ (W/m² K)
4. Calculation of U_{dirty} $U_{dirty} = \cfrac{1}{\frac{1}{U_{clean}} + f_i + f_a}$ (W/m² K)
5. Calculation of the fouling reserve of the new heat exchanger: $\sum f = \cfrac{U_{clean} - U_{eq}}{U_{clean} \times U_{req}}$

Example 8: Calculation of the overall heat transfer coefficient and the fouling reserve

Data: $\alpha_i = 620$ W/m² K $\alpha_o = 699$ W/m² K $s = 2$ mm $\lambda = 56$ W/m K
$d_i = 16$ mm $d_o = 20$ mm $f_i = f_a = 0.0002$ m² K/W $U_{req} = 230$ W/m² K

$$\alpha_{io} = 620 \times \frac{16}{20} = 496 \text{ W/m}^2 \text{ K}$$

$$U_{clean} = \cfrac{1}{\frac{1}{496} + \frac{1}{699} + \frac{0.002}{56}} = 287 \text{ W/m}^2 \text{ K}$$

$$U_{\text{dirty}} = \frac{1}{\frac{1}{287} + 0.0002 + 0.0002} = 257 \text{ W/m}^2 \text{ K}$$

$$\sum f = \frac{U_{\text{clean}} - U_{\text{eq}}}{U_{\text{clean}} \times U_{\text{req}}} = \frac{287 - 230}{287 \times 230} = 0.0009 \frac{\text{m}^2 \text{ K}}{\text{W}}$$

A clear fouling reserve of 0.0005 m² K/W is available.

Often it can be worthwhile for the selection or the design of a heat exchanger to calculate the overall heat transfer coefficient per meter of the heat exchanger and consequently determine the required total length. The corresponding equation for the calculation of the overall heat transfer coefficient U_L per meter tube is as follows:

$$U_L = \frac{\pi}{\frac{1}{\alpha_i \times d_i} + \frac{1}{2 \times \lambda} \times \ln \frac{d_o}{d_i} + \frac{1}{\alpha_o \times d_o} + \frac{f_i}{d_i} + \frac{f_a}{d_o}} \quad (\text{W/m K})$$

From U_L the required total length L_{req} is calculated for a certain heat load Q as follows:

$$Q = UL \times L_{\text{req}} \times \text{CMTD} \ (\text{W}) \quad L_{\text{req}} = \frac{Q}{U_L \times \text{CMTD}} \ (\text{m})$$

Example 9: Calculation of the overall heat transfer coefficient per meter heat exchanger length

Data: $\alpha_i = 620 \text{ W/m}^2 \text{ K}$ $\alpha_o = 699 \text{ W/m}^2 \text{ K}$ $s = 2 \text{ mm}$ $\lambda = 56 \text{ W/m K}$
$d_i = 16 \text{ mm}$ $d_o = 20 \text{ mm}$ $f_i = f_a = 0.0002 \text{ m}^2 \text{ K/W}$
$Q = 180{,}000 \text{ W}$ $\text{LMTD} = 51 \text{ K}$

$$U_L = \frac{\pi}{\frac{1}{620 \times 0.016} + \frac{1}{2 \times 56} \times \ln \frac{20}{16} + \frac{1}{699 \times 0.02} + \frac{0.0002}{0.016} + \frac{0.0002}{0.02}} = 15.96 \text{ W/m K}$$

$$L_{\text{req}} = \frac{180{,}000}{15.96 \times 51} = 221 \text{ m}$$

6.2 CALCULATION OF THE TEMPERATURE GRADIENT IN A HEAT EXCHANGER

The temperature fall in the individual overall heat transfer resistances can be calculated using the heat flux density q (W/m²):

$$q = U_{\text{dirty}} \times \Delta t_{\text{tot}} = \alpha_o \times \Delta t_o = \alpha_{io} \times \Delta t_i = \frac{\lambda}{s} \times \Delta t_{\text{wall}} = \frac{1}{f_i} \times \Delta t_{f_i}$$

$$= \frac{1}{f_a} \times \Delta t_{f_a} \ (\text{W/m}^2)$$

$$\Delta t_o = \frac{q}{\alpha_o} \quad \Delta t_i = \frac{q}{\alpha_{io}} \quad \Delta t_{\text{wall}} = q \times \frac{s}{\lambda} \quad \Delta t_{f_i} = q \times f_i \quad \Delta f_a = q \times f_a$$

$$\frac{1}{U_{\text{dirty}}} = \frac{1}{\alpha_o}$$

Example 10: Calculation of the temperature gradient in the overall heat transfer resistances

Data: $\alpha_o = 699$ W/m^2 K $\alpha_i = 620$ W/m^2 K $\lambda = 56$ W/m K

           $f_a = 0.0002$ $f_i = 0.0002$ $s = 2$ mm

           $\Delta t_{tot} = 20\,°C = $ LMTD $d_i = 20$ mm $d_o = 25$ mm

$$\alpha_{io} = 620 \times \frac{20}{25} = 496 \text{ W/m}^2 \text{ K}$$

$$\frac{1}{U_{dirty}} = \frac{1}{699} + \frac{1}{496} + \frac{0.002}{56} + 0.0002 + 0.0002 = 0.0039 \quad U_{dirty} = 257.6 \text{ W/m}^2 \text{ K}$$

$$q = U \times \Delta t_{tot} = 257.6 \times 20 = 5151 \text{ W/m}^2$$

$$\Delta t_o = \frac{q}{\alpha_o} = \frac{5151}{699} = 7.4\,°C \qquad\qquad \Delta t_i = \frac{q}{\alpha_{io}} = \frac{5151}{496} = 10.4\,°C$$

$$\Delta t_{wall} = q \times \frac{s}{\lambda} = 5151 \times \frac{0.002}{56} = 0.2\,°C$$

$$\Delta t_{f_i} = q \times f_i = 5151 \times 0.0002 = 1\,°C \quad \Delta t_{f_a} = q \times f_a = 5151 \times 0.0002 = 1\,°C$$

$$Check: \quad \Delta t_{tot} = 7.4 + 10.4 + 1 + 1 + 0.2 = 20\,°C$$

The calculated temperature fall is shown in Figure 6.3.

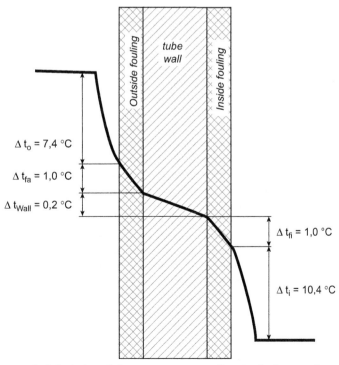

Figure 6.3 Calculation of the temperature gradient in the heat exchanger.

6.3 VISCOSITY CORRECTION

The viscosity has a predominant influence on the thickness of the laminar border film on the wall and this border layer determines the heat transfer coefficient. The higher the viscosity on the wall, thicker is the border film and worse the heat transfer coefficient.

The correction of the calculated heat transfer coefficient α_m for the average product temperature t_m follows from the viscosity ratio of the product at average fluid temperature and at wall temperature t_W.

$$\alpha_{korr} = \Phi \times \alpha_m$$

$$\text{Cooling}: \quad \frac{\eta_m}{\eta_W} < 1 \quad \Phi = \left(\frac{\eta_m}{\eta_W}\right)^{0.14}$$

$$\text{Heating}: \quad \frac{\eta_m}{\eta_W} > 1 \quad \Phi = \left(\frac{\eta_W}{\eta_m}\right)^{0.14}$$

α_{korr} = corrected heat transfer coefficient (W/m² K)
α_m = calculated α-value at t_m (W/m² K)
Φ = correction factor
η_m = viscosity at t_m (mPas)
η_W = viscosity at wall temperature t_W (mPas)

The wall temperature is smaller for cooling than the average product temperature and the viscosity increases. This deteriorates the heat transfer coefficient.

It is the opposite for heating. The lower viscosity on the wall improves the heat transfer coefficient.

Calculation procedure

1. Calculation of the heat transfer coefficient at average product temperature t_m

$$t_m = \frac{t_{in} + t_{out}}{2}$$

t_{in} = inlet temperature (°C)
t_{out} = outlet temperature (°C)

2. Determination of the wall temperature t_w

$$\text{Cooling}: \quad t_w = t_m - \Delta t \ (°C)$$
$$\text{Heating}: \quad t_w = t_m + \Delta t \ (°C)$$

Calculation of the temperature difference Δt between t_m and t_W (see Chapter 6.2)

$$\Delta t = \frac{q}{\alpha_m} \ (°C)$$

q = heat flux density (W/m²)

3. Derivation of the viscosity at the calculated wall temperature
4. Calculation of the viscosity correction Φ

Example 11: Calculation of the viscosity correction at thin media

Data: $\alpha_{io} = 336$ W/m² K $\alpha_o = 700$ W/m² K $U = 202.11$ W/m² K
          LMTD $= 16.83\,°C$ $q = 3401.92$ W/m²
Shell side: Water   25 °C → 30 °C $t_m = 27.5\,°C$

$$\Delta t_o = \frac{3401.92}{700} = 4.9\,°C$$

$$t_w = 27.5 + 4.9 = 32.4\,°C$$

The wall temperature is higher at the shell side than the average product temperature and the viscosity is smaller.

Tube side: Isopropanol   70 °C → 30 °C $t_m = 50\,°C$

$$\Delta t_i = \frac{3401.92}{336} = 10.1\,°C$$

$$t_w = 50 - 10.1 = 39.9\,°C$$

The wall temperature is smaller than the average product temperature and the viscosity on the wall is higher.

From Figure 6.4 the viscosities of water on the shell side and isopropanol in the tubes at different temperatures are taken.

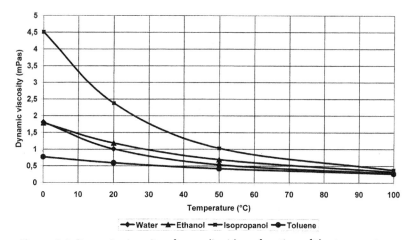

Figure 6.4 Dynamic viscosity of some liquids as function of the temperature.

Water: At 27.5 °C → $\eta_m = 0.89$ mPas   At 32.4 °C → $\eta_w = 0.8$ mPas

$$\left(\frac{\eta_m}{\eta_w}\right)^{0.14} = \left(\frac{0.89}{0.8}\right)^{0.14} = 1.01$$

$$\alpha_{korr} = 1.01 \times 336 = 339\ W/m^2\ K$$

The lower viscosity at the wall improves the heat transfer coefficient.

Isopropanol: At 50 °C ➡ $\eta_m = 1.08$ mPas   At 39.9 °C ➡ $\eta_w = 1.37$ mPas

$$\left(\frac{\eta_m}{\eta_W}\right)^{0.14} = \left(\frac{1.08}{1.37}\right)^{0.14} = 0.97$$

$$\alpha_{korr} = 0.97 \times 700 = 679\ W/m^2\ K$$

The higher viscosity on the wall worsens the heat transfer coefficient.

Example 12: Calculation of the viscosity correction for a high viscous oil

$\alpha_{io} = 50\ W/m^2\ K$ $\alpha_o = 1500\ W/m^2\ K$ $U = 47.15\ W/m^2\ K$ $t_m = 70\ °C$
LMTD $= 15\ °C$ $q = 707.229\ W/m^2$

$$\Delta t = \frac{707.229}{50} = 14.1\ °C \quad t_w = 70 - 14.1 = 55.9\ °C$$

$$\eta_m = 110\ \text{mPas at}\ 70\ °C \quad \eta_w = 192\ \text{mPas at}\ 55.9\ °C$$

$$\left(\frac{\eta_m}{\eta_W}\right)^{0.14} = \left(\frac{110}{192}\right)^{0.14} = 0.925 \quad \alpha_{korr} = 0.925 \times 50 = 46.2\ W/m^2 K$$

$$U = 43.8\ W/m^2\ K$$

Due to the high viscosity of the oil on the wall, the heat transfer coefficients and the overall heat transfer coefficients become worse.

6.4 CALCULATION OF THE HEAT TRANSFER COEFFICIENT FROM THE OVERALL HEAT TRANSFER COEFFICIENT

The following equation is valid for the overall heat transfer coefficient.

$$\frac{1}{U_{dirty}} = \frac{1}{\alpha_i} + \frac{1}{\alpha_o} + \frac{s}{\lambda} + f_i + f_a$$

After a simple conversion, the heat transfer coefficients can be determined.

$$\frac{1}{\alpha_i} = \frac{1}{U_{dirty}} - \frac{1}{\alpha_o} - \frac{s}{\lambda} - (f_i + f_a) \quad \frac{1}{\alpha_o} = \frac{1}{U_{dirty}} - \frac{1}{\alpha_i} - \frac{s}{\lambda} - (f_i + f_a)$$

Example 13: Determination of the α-values from the overall heat transfer coefficient U

Data: $U = 300\ \mathrm{W/m^2\ K}$ $\alpha_a = 1000\ \mathrm{W/m^2\ K}$ $f_i + f_a = 0.0004$ $s/\lambda = 0.00014$

$$\frac{1}{\alpha_i} = \frac{1}{300} - \frac{1}{1000} - 0.00014 - 0.0004 \quad \alpha_i = 557.6\ \mathrm{W/m^2\ K}$$

$$\frac{1}{\alpha_o} = \frac{1}{300} - \frac{1}{557.6} - 0.00014 - 0.0004 \quad \alpha_o = 1000\ \mathrm{W/m^2\ K}$$

Kontrolle

$$\frac{1}{U_{dirty}} = \frac{1}{\alpha_i} + \frac{1}{\alpha_o} + \frac{s}{\lambda} + f_i + f_a = \frac{1}{557.6} + \frac{1}{1000} + 0.00014 + 0.0004 \quad U = 300\ \mathrm{W/m^2\ K}$$

NOMENCLATURE

A area of the exchanger (m^2)
α_i tube-side heat transfer coefficient (W/m^2 K)
α_{io} α_i based on the outer area (W/m^2 K)
α_o shell-side heat transfer coefficient (W/m^2 K)
d_i tube inner diameter (m)
d_o tube outer diameter (m)
f_i inner fouling resistance (m^2 K/W)
f_a outer fouling resistance (m^2 K/W)
U general overall heat transfer coefficient (W/m^2 K)
U_α overall heat transfer coefficient calculated from both α-values (W/m^2 K)
U_{clean} overall heat transfer coefficient without fouling (W/m^2 K)
U_{dirty} overall heat transfer coefficient with fouling (W/m^2 K)
U_{req} required overall heat transfer coefficient (W/m^2 K)
U_{ex} overall heat transfer coefficient considering fouling + tube form (W/m^2 K)
U_L overall heat transfer coefficient per meter tube (W/m^2 K)
Λ heat conductivity of the tube material (W/m K)
L length of the heat exchanger tubes (m)
L_{req} required total length of the tubes (m)
CMTD corrected effective mean temperature difference (K)
Q heat load (W)
s tube wall thickness (m)

REFERENCES

[1] D.Q. Kern, Process Heat Transfer, McGraw-Hill, NY, 1950.
[2] W.H. McAdams, Heat Transmission, McGraw-Hill, NY, 1954.
[3] Perry's Chemical Engineer's Handbook, McGraw-Hill, 1984.
[4] G.F. Hewitt, G.L. Shires, T.R. Bott, Process Heat Transfer, CRC Press, Boca Raton, 1994.
[5] Lord, Minton, Slusser, Design of heat exchangers, Chem. Eng. 26 (Jan. 1970).

[6]   N.P. Chopey, Heat transfer, in: Handbook of Chemical Engineering Calculations, McGraw-Hill, New York, 1993.

[7]   V.D.I. Wärmeatlas, 5. Aufkage, VDI-Verlag, Düsseldorf, 1988.

[8]   F. Hell, Grundlagen der Wärmeübertragung, VDI-Verlag, Düsseldorf, 1982.

[9]   W.M. Rohsenow, J.P. Hartnett, Handbook of Heat Transfer, McGraw-Hill, NY, 1973.

[10]   E.R.G. Eckert, R.M. Drake, Heat and Mass Transfer, McGraw Hill, NY, 1959.

CHAPTER 7

Chemical Engineering Calculations

Contents

In the design of condensers and evaporators for multicomponent mixtures, the dew point and dew point line and the bubble point and bubble point line must be calculated.

In condensers the first droplet is condensed at the dew point of the mixture, whereas the mixture is totally liquid at the bubble point.

The vapor mixture must be cooled from the dew point to the bubble point.

In evaporators of liquid mixtures, the first droplet evaporates at the bubble point and at the dew point the total mixture is in vapor state.

The mixture must be heated from the bubble point to the dew point.

For the design of condensers and evaporators, the condensation, that is, flash curves and the heat load curve with the heat load requirement as a function of the temperature is needed (Figure 7.1).

7.1 VAPOR PRESSURE CALCULATIONS [5]

The vapor pressure is calculated using the Antoine equation and the Antoine constants A, B, and C which are available in the literature. Some vapor pressure curves are shown in Figure 7.2.

Antoine equation:

$$lg\, p_0 = A - \frac{B}{C + t(°C)} \text{ (mbar)}$$

Heat Exchanger Design Guide
http://dx.doi.org/10.1016/B978-0-12-803764-5.00007-9

© 2016 Elsevier Inc.
All rights reserved.

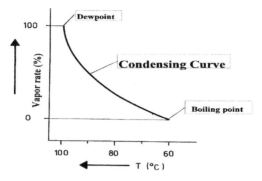

Figure 7.1 Condensing curve.

Example 1: Vapor pressure calculation for benzene and toluene

	Benzene	Toluene
	A = 7.00481	A = 7.07581
	B = 1196.76	B = 1342.31
	C = 219.161	C = 219.187

Temperature	Vapor pressures:	
50 °C:	p_{OB} = 360.4 mbar	p_{OT} = 122.5 mbar
98 °C:	p_{OB} = 1710 mbar	p_{OT} = 696 mbar
103 °C:	p_{OB} = 1960 mbar	p_{OT} = 810 mbar

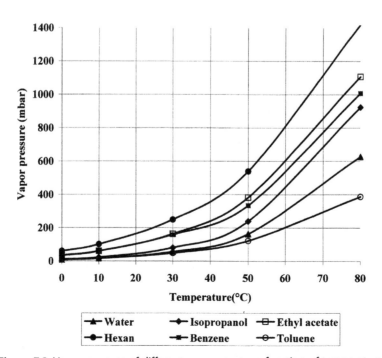

Figure 7.2 Vapor pressure of different components as function of temperature.

7.2 EQUILIBRIUM BETWEEN THE LIQUID AND THE VAPOR PHASE [5]

Dalton (Figure 7.3):

$$p_1 = y_1 \times P_{tot} \qquad P_{tot} = p_1 + p_2 + p_3 + ...p_i$$

Raoult (Figure 7.3):

$$p_1 = x_1 \times p_{01} \qquad p_2 = x_2 \times p_{02}$$

$$P_{tot} = x_1 \times p_{01} + x_2 \times p_{02} + x_3 \times p_{03} + ...x_i \times p_{0i}$$

Equilibrium equation: $y_i \times P_{tot} = p_i = x_i \times p_{0i}$
Calculation of the equilibrium factor K:

$$K = \frac{y_i}{x_i} = \frac{p_{0i}}{P_{tot}}$$

y_i = composition of the component i in the vapor phase (mole fraction)
x_i = composition of the component i in the liquid phase (mole fraction)
P_{tot} = total pressure (mbar)
p_{0i} = vapor pressure of the component i (mbar)
p_i = partial pressure of the component i (mbar)

Example 2: Calculation of the partial pressures according to Raoult and the vapor composition according to Dalton and the equilibrium factors

$$x_1 = 0.6 \qquad p_{01} = 800 \text{ mbar}$$
$$x_2 = 0.4 \qquad p_{02} = 1300 \text{ mbar}$$

Calculation of the partial pressures according to Raoult:

$$p_1 = x_1 \times p_{01} = 0.6 \times 800 = 480 \text{ mbar}$$
$$p_2 = x_2 \times p_{02} = 0.4 \times 1300 = 520 \text{ mbar}$$
$$P_{tot} = 1000 \text{ mbar}$$

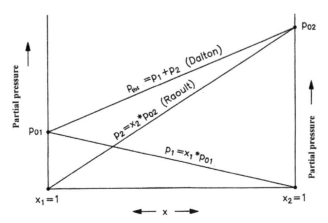

Figure 7.3 Graphical representation of the laws of Dalton and Raoult.

Calculation of the vapor composition according to Dalton:

$$y_1 = \frac{p_1}{P_{tot}} = \frac{480}{1000} = 0.48 = 48\ \text{Mol\%}$$

$$y_2 = \frac{p_2}{P_{tot}} = \frac{520}{1000} = 0.52 = 52\ \text{Mol\%}$$

Calculation of the equilibrium factors K:

$$K_1 = \frac{y_1}{x_1} = \frac{0.48}{0.6} = \frac{p_{01}}{P_{tot}} = \frac{800}{1000} = 0.8$$

$$K_2 = \frac{y_2}{x_2} = \frac{0.52}{0.4} = \frac{p_{02}}{P_{tot}} = \frac{1300}{1000} = 1.3$$

7.3 BUBBLE POINT CALCULATION [5]

The bubble point of a mixture is defined as follows:

$$\sum y_i = \sum K_i \times x_i = 1$$

The bubble point pressure P_{bub} is calculated directly.

$$P_{bub} = x_1 \times p_{01} + x_2 \times p_{02}$$

The bubble point temperature for a particular total pressure is determined iteratively by calculating the bubble point pressure at different temperatures.

The following example shows the procedure.

Example 3: Iterative bubble point calculation for a benzene—toluene mixture

30 Mol% benzene in the liquid phase ($x = 0.3$)
70 Mol% toluene in the liquid phase ($x = 0.7$)
$P_{tot} = 1000$ mbar

    1. Choice: $t = 95\ °C$ $p_{0B} = 1573$ mbar $p_{0T} = 634.4$ mbar
         $P_{bub} = 0.3 \times 1573 + 0.7 \times 634.4 = 916$ mbar $P_{bub} < P_{tot}$

    2. Choice: $t = 100\ °C$ $p_{0B} = 1807$ mbar $p_{0T} = 740$ mbar
         $P_{bub} = 0.3 \times 1807 + 0.7 \times 740 = 1060$ mbar $P_{bub} > P_{tot}$

    3. Choice: $t = 98\ °C$ $p_{0B} = 1710$ mbar $p_{0T} = 696$ mbar
         $P_{bub} = 0.3 \times 1710 + 0.7 \times 696 = 1000$ mbar $P_{bub} = P_{tot} = 1000$ mbar

Check calculation at 98 °C:

$$K_1 = \frac{p_{0B}}{P_{tot}} = \frac{1700}{1000} = 1.71$$

$$K_2 = \frac{p_{0T}}{P_{tot}} = \frac{696}{1000} = 0.696$$

$$\sum K_i \times x_i = 1.71 \times 0.3 + 0.696 \times 0.7 = 1$$

Thus, the bubble point specification is fulfilled.

7.4 DEW POINT CALCULATION [5]

The dew point of a vapor mixture is defined as follows:

$$\sum x_i = \sum y_i / K_i = 1$$

The dew point pressure P_{dew} is calculated directly.

$$\frac{1}{P_{dew}} = \frac{y_1}{p_{01}} + \frac{y_2}{p_{02}}$$

The dew point temperature for a particular total pressure is calculated iteratively. The following example shows the procedure.

Example 4: Iterative dew point calculation for a benzene–toluene mixture

30 Mol% benzene in the vapor phase ($y = 0.3$)
70 Mol% toluene in the vapor phase ($y = 0.7$)
$P_{tot} = 1000$ mbar

    1. Choice: $t = 100\ °C$ $p_{0B} = 1807$ mbar $p_{0T} = 740$ mbar

$$\frac{1}{P_{dew}} = \frac{0.3}{1807} + \frac{0.7}{740} = 0.0011 \quad P_{dew} = 899\ \text{mbar} < P_{tot} = 1000\ \text{mbar}$$

    2. Choice: $t = 104\ °C$ $p_{0B} = 2013$ mbar $p_{0T} = 834.4$ mbar

$$\frac{1}{P_{dew}} = \frac{0.3}{2013} + \frac{0.7}{834.4} = 0.00099 \quad P_{dew} = 1012\ \text{mbar} > P_{tot} = 1000\ \text{mbar}$$

    3. Choice: $t = 103.8\ °C$ $p_{0T} = 2002$ mbar $p_{0T} = 829$ mbar

$$\frac{1}{P_{dew}} = \frac{0.3}{2002} + \frac{0.7}{829} = 0.001 \quad P_{dew} = 1006\ \text{mbar} \approx P_{tot} = 1000\ \text{mbar}$$

Check calculation at 103.8 °C:

$$K_1 = \frac{2002}{1006} = 1.99 \qquad K_2 = \frac{829}{1006} = 0.824$$

$$\sum \frac{y_i}{K_i} = \frac{0.3}{1.99} + \frac{0.7}{0.824} = 1$$

The condition for the dew point is reached.

7.5 CALCULATION OF DEW AND BUBBLE LINES OF IDEAL BINARY MIXTURES [5]

The bubble point of a liquid mixture is defined such that the sum of the partial pressures of the mixture reaches the system pressure and the first droplet is evaporated.

Due to the preferential evaporation of light components, the heavier components in the liquid mixture increase and therefore the bubble point increases. The representation of the bubble point temperature as a function of the composition of a lighter component in the mixture is called the bubble line.

Equation of the bubble line:

$$x_1 = f(t) = \frac{P_{tot} - p_{02}}{p_{01} - p_{02}} \text{ (Molfraction of the light component in the liquid)}$$

The bubble point temperature increases with decreasing concentration of the light boiling components.

The dew point of a vapor mixture is the temperature in which the first droplet condenses. Due to the preferential condensation of the heavier component, the contents of the lighter components in the vapor mixture increase and the dew point temperature falls.

The representation of the dew point temperature as a function of the vapor compositions is known as the dew line.

Equation of the dew line:

$$y_1 = f(t) = \frac{p_{01}}{P_{tot}} \times \frac{P_{tot} - p_{02}}{p_{01} - p_{02}} \text{ (Molfraction of the light component in the vapor)}$$

A diagram with bubble and dew line is known as the phase diagram (Figure 7.4).

Example 5: Construction of the phase diagram for benzene (1)—toluene (2) at 1.013 bar

Procedure:

1. Calculation of both bubble points with the Antoine equation. The two bubble points are the boundary points in the phase diagram.

Boiling point of benzene: 80.1 °C   Boiling point of toluene: 110.6 °C

2. Calculation of the vapor pressures at different temperatures with the Antoine equation.
3. Determination of the liquid composition x and the vapor composition y at different temperatures.

Temperature (°C)	p_{01} (mbar)	p_{02} (mbar)	x_1	y_1
80.1	1013	390	1.000	1.000
83	1107	430	0.861	0.941
86	1211	476	0.731	0.874
89	1322	525	0.612	0.799
92	1442	578	0.504	0.717
95	1569	636	0.404	0.626
98	1705	698	0.313	0.527
101	1850	765	0.229	0.418
104	2004	836	0.151	0.300
107	2168	913	0.080	0.170
110.6	2378	1013	0.000	0.000

Bubble line: $x_1 = f$ (temperature)   Dew line: $y_1 = f$ (temperature)

In Figure 7.4 the phase diagram of the benzene–toluene mixture is presented.

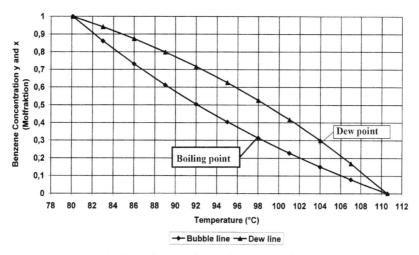

Figure 7.4 Phase diagram for the benzene–toluene mixture.

The dew point of a vapor mixture with 30 mol% benzene and 70 mol% toluene is around 104 °C.
The bubble point of the mixture is around 98 °C.
The first liquid droplet condenses at 104 °C at the dew point.
At the bubble point at 98 °C, the total mixture is liquid.
In order to condense the total mixture, the mixture must be cooled down from 104 °C to 98 °C.
On the contrary, in vaporizing the mixture must be heated from the bubble point to the dew point.

7.6 FLASH CALCULATIONS [1,2,5]

The vapor and liquid rates change in the region between the bubble and the dew point.

When condensing the vapor rate decreases as the temperature becomes lower.

When evaporating, more vapor is formed with increased heating.

With help of flash calculation the amount of vapor at temperatures between the bubble and the dew point is determined (Figure 7.5).

The compositions in the vapor and liquid phase when cooling or heating are also determined using the equilibrium factors (Figure 7.5).

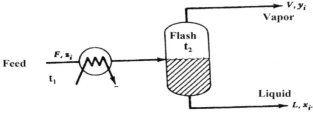

Figure 7.5 Flash separation.

At the bubble point, the total mixture is in the liquid phase.

At the dew point, the total mixture is in the vapor phase.

How much of the mixture is in the vapor phase at temperatures between the bubble point and the dew point? Figure 7.5.

The amount of vapor V in the feed F is calculated using the equilibrium constants K_1 and K_2 as follows:

$$\frac{V}{F} = \frac{z_1 \times \frac{K_1 - K_2}{1 - K_2} - 1}{K_1 - 1}$$

What are the compositions of the liquid and the vapor phase of the mixture?

$$x_1 = \frac{1 - K_2}{K_1 - K_2} \text{(Molfraction)} \qquad y_1 = K_1 \times x_1 \text{(Molfraction)}$$

V = vapor rate (kmol/h)

F = feed rate (kmol/h)

K_1 = equilibrium constant of the lighter component 1

K_2 = equilibrium constant of the heavier component 2

z_1 = composition of the lighter component 1 in feed (molfraction)

x_1 = composition of the lighter component in the liquid phase (molfraction)

y_1 = composition of the lighter component in vapor (molfraction)

Example 6: Flash calculations for a benzene–toluene mixture

At the bubble point (98 °C), all is liquid

At the dew point (104 °C), all is vapor

Feed composition: 30 mol% benzene in the vapor ($z_1 = 0.3$ mol fraction)

The amount in the vapor phase at 100 °C is to be determined

$t = 100\ °C$ $\qquad\qquad P_{tot} = 1000\ mbar$

Vapor pressure benzene $\qquad$ *Vapor pressure toluene*

$p_{0B} = 1807\ mbar$ $\qquad\qquad p_{0T} = 740\ mbar$

$P_{tot} = 1000\ mbar$

$$Benzene: \quad K_1 = \frac{p_{0B}}{P_{tot}} = \frac{1807}{1000} = 1.807$$

$$Toluene: \quad K_2 = \frac{p_{0T}}{P_{tot}} = \frac{740}{1000} = 0.74$$

$$\frac{V}{F} = \frac{0.3 \times \frac{1.807 - 0.74}{1 - 0.74} - 1}{1.807 - 1} = 0.2864$$

28.64 mol% of the feed is vapor!

Calculation of the compositions in liquid and vapor at $t = 100\ °C$

$$x_1 = \frac{1 - K_2}{K_1 - K_2} = \frac{1 - 0.74}{1.807 - 0.74} = 0.2437 \qquad \rightarrow 24.37\ mol\%\ benzene\ in\ the\ liquid$$

$$y_1 = K_1 \times x_1 = 1.807 \times 0.2437 = 0.44 \qquad \rightarrow 44\ mol\%\ benzene\ in\ the\ vapor$$

7.7 CONDENSATION OR FLASH CURVE OF BINARY MIXTURES

For the design of condensers and evaporators for mixtures, the condensation or flash curve is required.

In these curves, the vapor fraction V/F of the mixture based on the feed rate F is plotted over the temperature of the mixture.

The calculations follow the equations in Section 7.6.

Example 7: Calculation of the condensation curve for the mixture benzene-*o*-xylene

Inlet composition: $z_1 = 0.576 = 57.6$ mol% *benzene in the vapor mixture*

T (°C)	K_1	K_2	V/F	x_1	y_1
96	1.57	0.22	0	0.576	0.906
98.9	1.74	0.25	0.2	0.502	0.874
104	2	0.3	0.4	0.41	0.82
110.2	2.34	0.37	0.6	0.318	0.748
116.2	2.72	0.448	0.8	0.242	0.66
121	3.07	0.52	1	0.188	0.576

$T = 96\,°C$:
$$\frac{V}{F} = \frac{z_1 \times \dfrac{K_1 - K_2}{1 - K_2} - 1}{K_1 - 1} = \frac{0.576 \times \dfrac{1.57 - 0.22}{1 - 0.22} - 1}{1.57 - 1} = 0$$

$$x_1 = \frac{1 - K_2}{K_1 - K_2} = \frac{1 - 0.22}{1.57 - 0.22} = 0.576 \quad y_1 = K_1 \times x_1 = 1.57 \times 0.576 = 0.906$$

$T = 98.9\,°C$
$$\frac{V}{F} = \frac{0.576 \times \dfrac{1.74 - 0.25}{1 - 0.25} - 1}{1.74 - 1} = 0.2$$

$$x_1 = \frac{1 - 0.25}{1.74 - 0.25} = 0.502 \quad y_1 = 1.74 \times 0.502 = 0.874$$

$T = 116.2\,°C$
$$\frac{V}{F} = \frac{0.576 \times \dfrac{2.72 - 0.448}{1 - 0.448} - 1}{2.72 - 1} = 0.8$$

$$x_1 = \frac{1 - 0.448}{2.72 - 0.448} = 0.242 \quad y_1 = 2.72 \times 0.242 = 0.66$$

$T = 121\,°C$
$$\frac{V}{F} = \frac{0.576 \times \dfrac{3.07 - 0.52}{1 - 0.52} - 1}{3.07 - 1} = 1$$

$$x_1 = \frac{1 - 0.52}{3.07 - 0.52} = 0.188 \quad y_1 = 3.07 \times 0.188 = 0.576$$

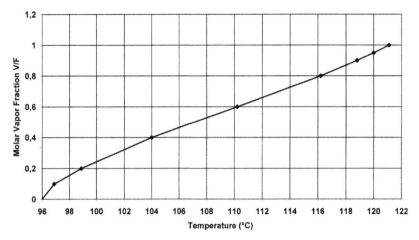

Figure 7.6 Condensation curve of the mixture benzene-*o*-xylene of Example 7.

The condensation curve for the mixture benzene-*o*-xylene is shown in Figure 7.6.

The molar vapor fraction V/F drops from $V/F = 1$ at the dew point at 121 °C to $V/F = 0$ at the bubble point 96 °C.

After conversion from mole to weight units, the vapor and condensate rate in kilogram per hour at different temperatures shown in Figure 7.7 results for a feed rate of 2000 kg/h with 50 weight% benzene and 50 weight% *o*-xylene.

The required heat loads for the condensation of the benzene-*o*-xylene mixture result from the enthalpies for cooling the vapor mixture, the condensation and the cooling of the condensate.

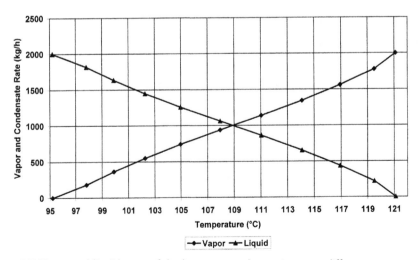

Figure 7.7 Vapor and liquid rates of the benzene-*o*-xylene mixture at different temperatures.

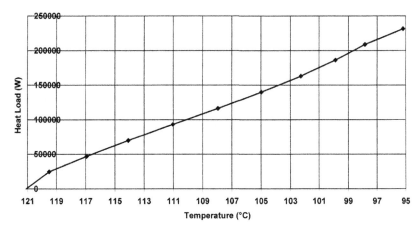

Figure 7.8 Heat load curve for the condensation of 2 t/h of the benzene-*o*-xylene mixture as a function of the temperature.

In Figure 7.8 the determined heat loads for the condensing of a mixture of 2000 kg/h benzene-*o*-xylene are shown as a function of the temperature.

In condensing and cooling from dew point to bubble point of the mixture, the required heat load rises from 0 to 230 kW.

7.8 CALCULATION OF NONIDEAL BINARY MIXTURES [3,4,5]

In the calculation of phase equilibriums of nonideal mixtures the activity coefficient γ must be considered.

$$P_{tot} = \gamma_1 \times x_1 \times p_{01} + \gamma_2 \times x_2 \times p_{02} \text{ (mbar)}$$

$$y_1 \times P_{tot} = \gamma_1 \times x_1 \times p_{01}$$

$$y_1 = \frac{x_1 \times p_{01} \times \gamma_1}{P_{tot}} \text{ (Molfraction)}$$

$$K_1 = \frac{p_{01} \times \gamma_1}{P_{tot}} \qquad K_2 = \frac{p_{02} \times \gamma_2}{P_{tot}}$$

The activity coefficient γ is strongly dependent on the liquid composition and can be calculated with different models: Wilson, NRTL, Uniquac, Unifac [3,5]

Example 8: Bubble point calculation with activity coefficient for the mixture methanol (1)—water (2)

$x_1 = 0.1$	$x_2 = 0.9$	$t = 87.8\,°C$
$\gamma_1 = 1.705$	$\gamma_2 = 1$	
$p_{01} = 2444$ mbar	$p_{02} = 648$ mbar	

$$P_{bub} = x_1 \times \gamma_1 \times p_{01} + x_2 \times \gamma_2 \times p_{02} = 0.1 \times 1.705 \times 1833 + 0.9 \times 1 \times 486 = 1000 \text{ mbar}$$

$$K_1 = \frac{\gamma_1 \times p_{01}}{p_{tot}} = \frac{1.705 \times 2444}{1000} = 4.167$$

$$K_2 = \frac{1 \times 648}{1000} = 0.64$$

Bubble point check: $\Sigma \, K_i \times x_i = 0.1 \times 4.167 + 0.9 \times 0.64 = 1$

The bubble point condition is reached at 87.8 °C!

Without considering the activity coefficient γ the bubble point temperature of 93.1 °C instead of 87.8 °C results.

Example 9: Dew point calculation with activity coefficient γ for the mixture methanol (1)—water (2)

$y_1 = 0.1$ $y_2 = 0.9$ $t = 97.1$ °C

$\gamma_1 = 2.2$ $\gamma_2 = 1.0$

$p_{01} = 3343$ mbar $p_{02} = 910$ mbar

$$\frac{1}{P_{dew}} = \frac{y_1}{\gamma_1 \times p_{01}} + \frac{y_2}{\gamma_2 \times p_{02}} = \frac{0.1}{2.2 \times 3343} + \frac{0.9}{910} = 0.001$$

$$P_{dew} = 1 \text{ bar}$$

$$K_1 = \frac{2.2 \times 3343}{1000} = 7.35 \qquad K_2 = \frac{1 \times 910}{1000} = 0.91$$

$$\sum \frac{y_i}{K_i} = \frac{0.1}{7.35} + \frac{0.9}{0.91} = 1$$

The dew point condition is reached at 97.1 °C!

Example 10: Flash calculation for the mixture methanol (1)—water (2)

$P = 1$ bar $= 1000$ mbar $t = 92.1$ °C   Feed rate $= 100$ kMol/h

Inlet vapor compositions: $z_1 = 0.1$ $z_2 = 0.9$

       $\gamma_1 = 1.848$ $\gamma_2 = 1.004$

       $p_{01} = 2817$ mbar $p_{02} = 756$ mbar

$$K_1 = \frac{1.848 \times 2817}{1000} = 5.2 \qquad K_2 = \frac{1.004 \times 756}{1000} = 0.76$$

$$\frac{V}{F} = \frac{z_1 \times \frac{K_1 - K_2}{1 - K_2} - 1}{K_1 - 1} = \frac{0.1 \times \frac{5.2 - 0.76}{1 - 0.76} - 1}{5.2 - 1} = 0.20$$

At 92.1 °C 20% of the feed are vapor.

Vapor rate $V = 0.20 \times 100 = 20$ kmol/h

Liquid rate $L = 100 - 20 = 80$ kmol/h

Calculation of the composition in the liquid and in the vapor:

$$x_1 = \frac{1 - K_2}{K_1 - K_2} = \frac{1 - 0.76}{5.2 - 0.76} = 0.054 \text{ Molfraction}$$

$x_2 = 1 - 0.054 = 0.946$ (molfraction) $= 94.6$ Mol%
$y_1 = K_1 \times x_1 = 5.2 \times 0.054 = 0.28$ (molfraction) $= 28$ mol%
$y_2 = K_2 \times x_2 = 0.76 \times 0.946 = 0.72$ (molfraction) $= 72$ mol%
Without considering the activity coefficient γ, the calculation gives the following wrong results at 92.1 °C:

$$K_1 = 2.81 \qquad\qquad K_2 = 0.756$$
$$V = 0 \qquad\qquad L = 100 \text{ kmol/h}$$

Nothing is vaporized because the bubble point lies with $\gamma = 1$ at 93.1 °C, that is, over 92.1 °C.

7.9 FLASH CALCULATIONS FOR MULTICOMPONENT MIXTURES [1,2]

The calculation must be performed iteratively for mixtures with more than two components.

The individual value for V/F of the different components is calculated with an estimated value for V/F. The sum of the V/F values of the components must equal the V/F estimated value.

$$\frac{V}{F} = \sum \frac{z_i}{1 + \dfrac{L}{V \times K_i}} \qquad\qquad \frac{L}{V} = \frac{F}{V} - 1 = \frac{1}{V/L}$$

$$y_i = \frac{F}{V} \times \left(\frac{z_i}{1 + \dfrac{L}{V \times K_i}} \right) \qquad\qquad x_i = \frac{F}{V} \times \left(\frac{z_i}{K_i + \dfrac{L}{V}} \right)$$

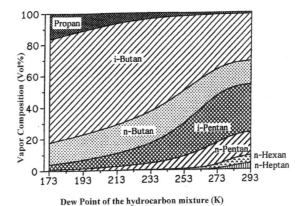

Figure 7.9 Vapor composition of a hydrocarbon mixture as function of the temperature.

A description of the procedure with example is found in references (1) and (2).

From Figure 7.9, it can be seen that the dew point temperature greatly falls with increasing concentration of the low-boiling components.

REFERENCES

[1] E.J. Henley, J.D. Seader, Equilibrium-stage Separation Operations in Chemical Engineering, John Wiley, New York, 1981.
[2] B.D. Smith, Design of Equilibrium Stage Processes, McGraw-Hill, New York, 1963.
[3] J. Gmehling, B. Kolbe, M. Kleiber, J. Rarey, Chemical Thermodynamics for Process Simulation, Wiley-VCH-Verlag, Weinheim, 2012.
[4] G. Mehos, Estimate binary equilibrium coefficients, Chem. Eng. 103 (1996) 131.
[5] M. Nitsche, Kolonnen-fibel, Springer Verlag, Berlin, 2014.

CHAPTER 8

Design of Condensers

Contents

Media in vapor state are condensed on cold surfaces if the cooling surface is colder than the dew point of the vapors. In film condensation a condensate film forms on the cooling area.

In technical plants the vapors are liquefied by film condensation.

In isothermal condensation of single components, condensation occurs at the dew point.

In multicomponent mixtures, in condensing, the mixture must be cooled and liquefied according to the condensation line from the dew point to the bubble point.

Heat Exchanger Design Guide
http://dx.doi.org/10.1016/B978-0-12-803764-5.00008-0

© 2016 Elsevier Inc.
All rights reserved.

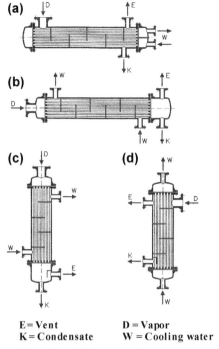

E = Vent   D = Vapor
K = Condensate   W = Cooling water

Figure 8.1 Different condenser types.

CONSTRUCTION TYPES OF CONDENSERS

The following indirect condensation types are distinguished:

Condensation on horizontal tube bundle (Figure 8.1(a))

Condensation in horizontal tubes (Figure 8.1(b))

Condensation in vertical tubes (Figure 8.1(c))

Condensation on vertical tube bundle (Figure 8.1(d))

The four principal different condenser types are shown in Figure 8.1.

In Figure 8.2 it can be seen how the heat transfer coefficients in the different condenser types differ in the condensation of ethanol and cyclohexane and how the heat transfer coefficients change with increasing condensate rate.

The heat transfer coefficients are best in the case of horizontal condensation.

With increasing condensate rate or growing condensate film thickness the heat transfer coefficients fall. When condensation takes place in the tubes the α-value becomes better for higher vapor loadings because the condensate film becomes turbulent and the vapor shear stress improves the heat transfer.

8.1 CONDENSER CONSTRUCTION TYPES

In the following the operation of the different condenser construction types shown in Figure 8.4 is described.

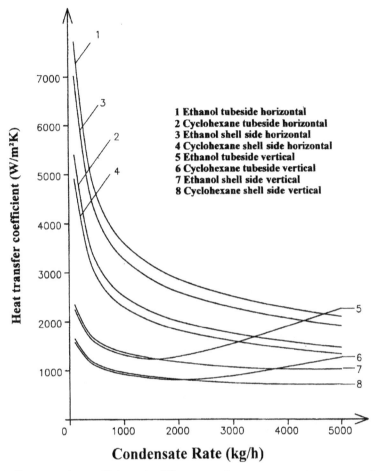

Figure 8.2 Heat transfer coefficients in different condenser construction types as function of the condensate rate.

8.1.1 Condensation in the shell chamber on the horizontal tube bundle (Figure 8.4(a))

The vapors enter from the top and flow through a horizontal tube bundle.

Cooling water flows through the tubes. The vapors condense on the cold tubes and the condensate runs over the tubes down to the bottom.

Due to the small condensate film thickness, good heat transfer coefficients result which reduce with increasing condensate rate.

Advantages: Low subcooling and more adequate for the condensation in vacuum because of the low pressure drop and the lack of an adiabatic mass flux capacity limit by a sudden cross-sectional constriction.

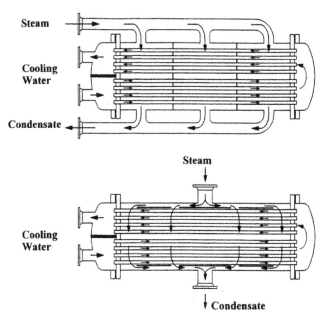

Figure 8.3 Low pressure drop cross condenser for the vacuum condensation.

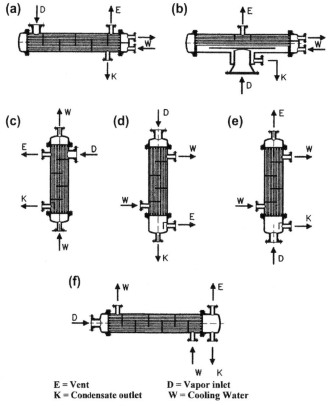

E = Vent   D = Vapor inlet
K = Condensate outlet W = Cooling Water

Figure 8.4 Different condenser constructions.

Tube pitch and baffle spacings can be adjusted to the vapor stream.

Low pressure loss and easy tube-side cleaning.

Disadvantages: Noncondensable gases must be removed, otherwise one part of the cooling area is covered. With a wide boiling mixture a differential condensation results. The high-boiling components are initially condensed, run down, and are no longer available for an equilibrium condensation. Hence the condensation of the concentrated low-boiling components becomes more difficult, for instance, in the condensation of a water–ammonia mixture.

It comes to a condensate congestion if no free flow area is provided at the bottom in the baffles. The baffles should be best arranged sideways.

In Figure 8.3, a cross stream condenser with low pressure drop for the condensation in vacuum is shown.

8.1.2 Condensation on horizontal water-cooled tubes in a top condenser (Figure 8.4(b))

The vapors flow through the shell chamber from the bottom to the top and are condensed on the horizontal tubes through which cooling water is flowing. To avoid short circuit flows a separating baffle plate must be installed. The condensate running from the tubes is collected in the bottom and is returned to the column as reflux or taken out as distillate product.

Advantages: Low holdup in distillate cycle and therefore absolutely necessary in batch distillations. High-boiling components are washed back by the later condensing light-boiling components.

Simple construction with the inert gas outlet on the top outmost and coldest point. Dephlegmator effect by partial condensation of multicomponent mixtures is possible.

Disadvantages: The vapors running upward are flowing against the condensed liquid and can therefore cause evaporation of light boiling components in the bottom region.

The light-boiling components accumulate toward the top and are then more difficult to condense because instead of an equilibrium a differential condensation occurs. It must be cooled deeper. If the vapor velocities are high, liquid entrainment may occur.

8.1.3 Condensation shell side on a vertical bundle (Figure 8.4(c))

The vapors are liquefied on the shell side outside the vertical tubes.

Advantages: This arrangement is required for a vertical thermosiphon evaporator with the two-phase flow in the tubes and the steam heating on the shell side.

Disadvantages: Noncondensable gases cannot be easily removed and therefore cover a part of the heat exchanger surface area. This causes a CO_2 corrosion in steam-heated evaporators.

Segmental baffles cause condensate buildup and constitute trap bags for inert gas. Disc baffles are more adequate.

8.1.4 Condensation in vertical tubes (Figure 8.4(d))

The vapors and the condensate flow in cocurrent direction from the top to the bottom through the tubes which are cooled in the countercurrent direction on the shell side.

Advantages: Countercurrent arrangement with maximum effective temperature difference.

Particularly adequate for condensate subcooling and optimal for avoiding differential condensation of wide boiling mixtures and streams containing inert gases.

The vapors enter the tubes with high velocity so that the heat transfer is improved by the shear stress, however this occurs at the cost of the pressure loss.

Due to the turbulence in the condensate film the heat transfer coefficient with Reynolds numbers over 1800 will be clearly better than with the laminar film condensation according to Nusselt.

Disadvantages: Not suitable as vacuum condenser because of the limited adiabatic mass flow capacity (Section 8.6.3) and not suitable as top condenser because of the required construction height for gravity reflux.

The inevitable subcooling of the reflux is undesirable, because then the first distillation stages in the column must operate as direct heat exchanger.

Sufficient baffles must be installed on the shell side in order to enable a cooling water velocity of 0.8–1.2 m/s.

8.1.5 Vertical countercurrent condensation in reflux coolers on stirred tanks (Figure 8.4(e))

The vapors flow from the bottom to the top through the tubes and the condensate runs back from the top to the bottom. The cooling water flows through the shell chamber.

Advantage: Greatly suitable as reflux cooler on stirred tanks.

Disadvantage: Danger of differential condensation because the high-boiling components are condensed out at the bottom.

A check calculation must be performed in order to avoid flooding when the outflowing condensate is blocked by the upward flowing vapors (see Section 8.6.7).

8.1.6 Condensation in horizontal tubes (Figure 8.4(f))

The vapors are condensed while flowing through the horizontal tubes.

The cooling water flows in countercurrent direction on the shell side.

Advantages: Very good heat transfer coefficients because of the thin condensate film thickness and the improvement by the shear stress. Also adequate as top condenser or air cooler.

Integral or equilibrium condensation because of the equilibrium between vapor and condensate.

Disadvantages: Not adequate for vacuum condensation because of the limited adiabatic mass flow capacity (Section 8.6.3).

Subcooling of the condensate cannot be avoided. The collected condensate reduces the cooling surface. Flooding of some tubes can occur in the case of sudden load variations (see Section 8.6.6).

Noncondensable gases cover a part of the tube surface.

Liquid drops can be entrained by the inert gases in the outlet accumulator if the flow cross-section for the gas is set too low.

8.2 HEAT TRANSFER COEFFICIENTS IN ISOTHERMAL CONDENSATION [1–6,8,9]

8.2.1 Calculation of the heat transfer coefficients

Using the equations listed in Tables 8.1 and 8.2 the heat transfer coefficients for the isothermal condensation can be easily calculated.

Owing to the fact that the condensate film becomes thicker with increasing condensation and the heat transfer coefficient decreases with increasing condensate film thickness, a condenser should be calculated zone wise. In common practice the division is in 10 zones.

Lumped calculations give an average condensate film thickness if the *condensate load* M_{liq} *is set to 50% of the total vapor rate W.*

The application of the calculation equations is shown in the Example 3.

In the *condensation in horizontal tubes* the gravity-driven heat transfer coefficient decreases with increasing vapor rate because the condensate film gets thicker.

On the other hand the shear stress-driven heat transfer coefficient increases with increasing vapor rate. This is shown in Figure 8.5.

The shear stress-driven heat transfer coefficient decreases with decreasing vapor rate. The gravity-driven heat transfer coefficient decreases with increasing condensate rate.

The condenser becomes flooded if the condensate loads in horizontal tubes are too high.

In order to avoid this, the condensate height h in the tube at the outlet should not exceed a fourth of the tube diameter d_i.

Table 8.1 Equations for the calculation of the heat transfer coefficients of isothermal condensation according to the Nitsche method

Gravity-driven heat transfer coefficient with laminar condensate film:

Equation 1.1: $\alpha_g = 1.47 \times \frac{\lambda}{L} \times \left(\frac{1}{Re_{liq}}\right)^{1/3} = 1.47 \times \lambda \times \left(\frac{g}{\nu^2} \times \frac{1}{Re_{liq}}\right)^{1/3}$ (W/m² K)

For lumped calculations the Reynolds number Re is calculated with $M_{liq} = \frac{W}{2}$ (kg/h)

Horizontal in tubes	Horizontal at tube bundle
$Re_{liq} = \frac{M_{liq} \times 4}{1800 \times n \times l_{tube} \times \eta_{liq}}$	$Re_{liq} = \frac{M_{liq} \times 4}{3600 \times n^{0.75} \times l_{tube} \times \eta_{liq}}$
Vertical in tubes	Vertical at tube bundle
$Re_{liq} = \frac{M_{liq} \times 4}{3600 \times n \times \pi \times d_i \times \eta_{liq}}$	$Re_{liq} = \frac{M_{liq} \times 4}{3600 \times n \times \pi \times d_o \times \eta_{liq}}$

Heat transfer coefficient for vertical condensation with wavy or turbulent condensate film:
Wave formation in range: 40 < Re > 1600

Equation 1.2: $\alpha_w = 1.007 \times \frac{\lambda}{L} \times \left(\frac{1}{Re_{liq}}\right)^{2/9}$ (W/m² K)

Turbulent condensate film from Re = 1600

Equation 1.3: $\alpha_t = 0.01725 \times \frac{\lambda}{L} \times \frac{Re_{liq}}{(Re_{liq}^{0.75} - 235) \times Pr^{-0.5} + 159}$ (W/m² K)

Heat transfer coefficient for the shear stress-controlled condensation in the tubes
Horizontal in the tubes
Equation 1.4

$$\alpha_\tau = 0.024 \times \frac{\lambda}{d_i} \times \left(\frac{G_{tot} \times d_i}{\eta_{liq}}\right)^{0.8} \times Pr^{0.43} \times \left(\frac{1 + \left(\frac{\rho_{liq}}{\rho_V}\right)^{0.5}}{2}\right) \text{ (W/m}^2 \text{ K)}$$

$G_{tot} = \frac{4 \times W}{3600 \times n \times \pi \times d_i^2}$ (kg/m² s) Chosen : $\alpha = \alpha_\tau$ if $\frac{\alpha_\tau}{\alpha_g} > 1.5$

If the *quotient* α_τ / α_g is higher than 1.5, α_τ is valid for the calculation
Vertical in the tubes
Equation 1.5

$$\alpha_\tau = 0.0099 \times \frac{\lambda}{d_i} \times \left(\frac{G_m \times d_i}{\eta_{liq}}\right)^{0.9} \times Pr^{0.5} \times \left(\frac{\rho_{liq}}{\rho_V}\right)^{0.5} \times \left(\frac{\eta_V}{\eta_{liq}}\right)^{0.1}$$

$G_m = \sqrt{\frac{W^2}{3}} \times \frac{4}{3600 \times n \times \pi \times d_i^2}$ (kg/m² s)

Chosen: $\alpha_\tau > \alpha_g \Rightarrow \alpha = \alpha_\tau$
If α_τ is greater than α_g, α_τ is valid for the calculation.

Table 8.2 Kern equations for the calculation of the heat transfer coefficients [1]

Heat transfer coefficient α_g for gravitation-driven isothermal condensation:

Equation 2.1: $\alpha_g = 0.925 \times \lambda \times \frac{\rho^2 \times g}{\eta \times G}$ (W/m^2 K)

G = condensate loading per unit periphery (kg/m s) with $M_{liq} = W/2$

Vertical in the tubes	Vertical outside tubes
$G_{Vi} = \frac{M_{liq}}{n \times \pi \times d_i \times 3600}$ (kg/m s)	$G_{Vo} = \frac{M_{liq}}{n \times \pi \times d_o \times 3600}$ (kg/m s)
Horizontal in the tubes	Horizontal outside tubes
$G_{Hi} = \frac{M_{liq}}{0.5 \times n \times l_{tube} \times 3600}$ (kg/m s)	$G_{Hi} = \frac{M_{liq}}{n^{0.66} \times l_{tube} \times 3600}$ (kg/m s)

Turbulence correction factor f_C for wavy vertical condensation at Re > 40:

Equation 2.2: $f_C = 0.8 \times \left(\frac{Re}{4}\right)^{0.11} = 0.8 \times \left(\frac{G_V}{\eta}\right)^{0.11}$ $\alpha_w = f_C \times \alpha_g$ (W/m^2 K)

Turbulent vertical condensation at Re > 2100 with equation 2.3:

$Re = \frac{4 \times G}{\eta} > 2100$ $\alpha_t = 0.0076 \times \left(\frac{\lambda^3 \times \rho^2 \times g}{\eta^2}\right)^{1/3} \times \left(\frac{4 \times G_V}{\eta}\right)^{0.4}$ (W/m^2 K)

Characteristic length $L = \left(\frac{\nu^2}{g}\right)^{1/3}$

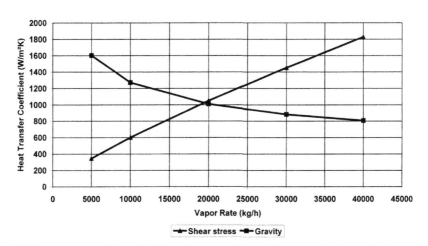

Figure 8.5 Shear stress- and gravity-driven heat transfer coefficients in horizontal tubes as function of vapor rate.

For the condition $h/d_i = 0.25$ the allowable condensate loading W_{allow} according to Section 8.6.6 is determined as follows [7]:

$$W_{allow} = 1570 \times n \times \rho \times d_i^{2.56} \text{ (kg/h)}$$

Example 1: Calculation of the allowable condensate load in horizontal tubes

Density $\rho = 614.3$ kg/m³

Condenser with 186 tubes 25 × 2, $d_i = 21$ mm, $L = 4$ m, $A = 58$ m²

$W_{allow} = 1570 \times 186 \times 614.3 \times 0.021^{2.56} = 9092$ kg/h

Condenser with 80 tubes 57 × 3.5, $d_i = 50$ mm, $L = 4$ m, $A = 57.3$ m²

$W_{allow} = 1570 \times 80 \times 614.3 \times 0.05^{2.56} = 36,036$ kg/h

The condenser capacity can be increased with larger tube diameters.

In *vertical tubes* the heat transfer becomes better if the condensate stream gets wavy or turbulent.

At high vapor velocities the heat transfer is highly improved by the shear stress.

This is shown in Figure 8.6.

The heat transfer coefficient reduces with increasing vapor rate because of the thicker condensate film. For wavy streams the heat transfer coefficient falls less greater and the heat transfer coefficient increases with increasing condensate rate for turbulent condensate stream.

Very good heat transfer coefficients are achieved with high vapor velocities due to the shear stress.

As a criterion for the improvement of the heat transfer by the shear force of the vapor flow the Wallis factor J is used [10].

$$J = \frac{m}{\sqrt{g \times d \times \rho_V \times (\rho_{liq} - \rho_V)}}$$

An improvement of the heat transfer by the shear stress is achieved at $J > 1.5$.

m = mass stream density (kg/m² s) = $\rho_V \times w_V$

ρ_V = vapor density (kg/m³)

ρ_{liq} = liquid density (kg/m³)

d = tube diameter (m)

w_V = vapor stream flow velocity (m/s)

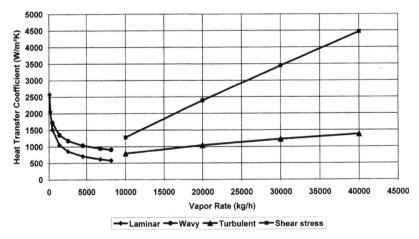

Figure 8.6 Heat transfer coefficients in vertical tubes as function of the vapor rate.

The mass stream density m is calculated as follows:

In the tubes:

$$m = \frac{W}{3600 \times n \times (\pi/4) \times d_i^2} = w_V \times \rho_V \, (kg/m^2 \, s)$$

On the shell side:

$$m = \frac{W}{B \times D_i \times \left(1 - \frac{d_o}{T}\right)} \, (kg/m^2 \, s)$$

B = baffle spacing (m)
D_i = shell inner diameter (m)
d_i = inner diameter (m)
d_o = outer diameter (m)
n = number of tubes
W = vapor rate (kg/h)

The condensation on the shell side of horizontal and vertical tube bundles is mostly gravity driven. The improvement of the heat transfer coefficient by the vapor shear stress in the first tube rows occur only at mass stream densities >30 kg/m^2 s.

Example 2: Calculation of the mass stream density m and the Wallis factor J

Data: Shell diameter $D_i = 0.55$ m   186 tubes 25 × 2 $L = 4$ m
Pitch $T = 32$ mm triangular   Baffle spacing $B = 0.94$ m
$\rho_D = 3.19$ kg/m^3 $\rho_F = 614$ kg/m^3 Vapor rate $W = 10,000$ kg/h

Shell side calculation:

$$m = \frac{W}{3600 \times B \times D_i \times \left(1 - \frac{d_a}{T}\right)} = \frac{10,000}{3600 \times 0.94 \times 0.55 \times \left(1 - \frac{25}{32}\right)} = 24.56 \, kg/m^2 \, s$$

$$\text{Shell side cross-section } f_c = B \times D_i \times \left(1 - \frac{d_a}{T}\right) = 0.94 \times 0.55 \times \left(1 - \frac{25}{32}\right) = 0.113 \, m^2$$

$$w_V = \frac{W}{3600 \times \rho_D \times f_q} = \frac{10,000}{3600 \times 3.19 \times 0.113} = 7.7 \, m/s$$

$$m = w_V \times \rho_V = 7.7 \times 3.19 = 24.56 \, kg/m^2 \, s$$

$$J = \frac{m}{\sqrt{g \times d \times \rho_V \times (\rho_{liq} - \rho_V)}} = \frac{24.56}{\sqrt{9.81 \times 0.025 \times 3.19 \times (614 - 3.19)}} = 1.12$$

Tube-side calculation:

$$m = \frac{W}{3600 \times n \times 0.785 \times d_i^2} = \frac{10,000}{3600 \times 186 \times 0.785 \times 0.021^2} = 43.1 \text{kg/m}^2 \text{ s}$$

Tube-side cross-section $f_T = n \times d_i^2 \times (\pi/4) = 186 \times 0.021^2 \times 0.785 = 0.06439 \text{ m}^2$

$$w_V = \frac{10,000}{3600 \times 3.19 \times 0.06439} = 13.52 \text{ m/s}$$

$$m = w_V \times \rho_V = 13.52 \times 3.19 = 43.1 \text{ kg/m}^2\text{s}$$

$$J = \frac{43.1}{\sqrt{9.81 \times 0.021 \times 3.19 \times (614 - 3.19)}} = 2.15$$

Example 3: Calculations of the heat transfer coefficients according to the equations listed in Tables 8.1 and 8.2

Heat exchanger data:	$A = 11.7 \text{ m}^2$	$D = 0.28 \text{ m}$	50 tubes 25 × 2,	$l_{tube} = 3 \text{ m}$
Physical data:	Vapor		Condensate	
Density (kg/m^3)	2.897		720	
Dynamic viscosity (Pas)	8.28×10^{-6}		0.000415 (0.415 mPas)	
Kinematic viscosity (m^2/s)	2.85×10^{-6}		0.576×10^{-6} (0.576 mm^2/s)	
Vaporization enthalpy (kJ/kg)	358.1		358.1	
Heat conductivity (W/m × K)	–		0.106	
Specific heat (kJ/kg K)	–		2.1567	
Prandtl number	–		8.44	
Characteristic length (m)	–		32.35×10^{-6}	

Total vapor rate $W = 2000$ kg/h cyclohexane

Set for lumped calculation with the average condensate rate

$M_{liq} = 0.5 \times 2000 = 1000$ kg/h condensate

Notice: The examples are calculated according to the lumped method, i.e., to the average of the condensate rate.

For comparison the results of the zone calculation are also listed in the result comparison.

8.2.1.1 Condensation at a horizontal tube bundle

Equation 1.1:

$$M_{fl} = 0.5 \times 2000 = 100 \text{ kg/h}$$

$$\text{Re}_{liq} = \frac{1000 \times 4}{3600 \times 50^{0.75} \times 3 \times 0.415 \times 10^{-3}} = 47.46$$

$$\alpha_g = 1.47 \times \frac{0.106}{32.35 \times 10^{-6}} \times \left(\frac{1}{47.46}\right)^{1/3} = 1330 \text{W/m}^2 \text{ K}$$

Kern equation 2.1:

$$W = 0.5 \times 2000 = 1000\,\text{kg/h} = M_{f1}$$

$$G_{ho} = \frac{1000}{50^{0.66} \times 3 \times 3600} = 0.007\,\text{kg/m s}$$

$$\alpha_g = 0.925 \times 0.106 \times \left(\frac{720^2 \times 9.81}{0.415 \times 10^{-3} \times 0.007}\right)^{1/3} = 1182\,\text{W/m}^2\,\text{K}$$

8.2.1.2 Condensation at a vertical tube bundle

Equation 1.1:

$$M_{liq} = 1000\,\text{kg/h}$$

$$Re_{liq} = \frac{1000 \times 4}{3600 \times 50 \times \pi \times 0.025 \times 0.415 \times 10^{-3}} = 681.8$$

$$\alpha_g = 1.47 \times \frac{0.106}{32.35 \times 10^{-6}} \times 681.8^{-1/3} = 547.5\,\text{W/m}^2\,\text{K}$$

Since the Reynolds number $Re_{fl} = 681$ is higher than 40 the *Equation 1.2 for wavy condensate flow must be used.*

Equation 1.2:

$$\alpha_w = 1.007 \times \frac{0.106}{32.35 \times 10^{-6}} \times 681.8^{-2/9} = 774\,\text{W/m}^2\,\text{K}$$

The heat transfer coefficient for wavy flow according to Equation 1.2 is higher.

Kern equation 2.1:

$$W = 1000\,\text{kg/h} = M_{fl}$$

$$G_{Vo} = \frac{1000}{50 \times \pi \times 0.025 \times 3600} = 0.0707\,\text{kg/m s}$$

$$\alpha_g = 0.925 \times 0.106 \times \left(\frac{720^2 \times 9.81}{0.415 \times 10^{-3} \times 0.0707}\right)^{1/3} = 547\,\text{W/m}^2\,\text{K}$$

Controlling the Reynolds number.

$$Re = \frac{4 \times 0.0707}{0.415 \times 10^{-3}} = 641 > Re = 40$$

For $Re > 40$ the turbulence correction is recommended according to *Equation 2.2:*

$$f_c = 0.8 \times \left(\frac{Re}{4}\right)^{0.11} = 0.8 \times \left(\frac{681}{4}\right)^{0.11} = 1.408$$

$$\alpha_w = 1.408 \times 547 = 770\,\text{W/m}^2\,\text{K}$$

8.2.1.3 Condensation in horizontal tubes

Equation 1.1:

$$M_{\text{liq}} = 1000 \text{ kg/h}$$

$$Re_{\text{liq}} = \frac{1000 \times 4}{1800 \times 50 \times 3 \times 0.415 \times 10^{-3}} = 35.7$$

$$\alpha_g = 1.47 \times \frac{0.106}{31.35 \times 10^{-6}} \times \left(\frac{1}{35.7}\right)^{1/3} \text{ W/m}^2 \text{ K}$$

$$\alpha_g = 1.47 \times 0.106 \times \left(\frac{9.81}{(0.576 \times 10^{-6})^2 \times 35.7}\right)^{1/3} = 1463 \text{ W/m}^2 \text{ K}$$

Calculation according to Equation 1.4 with vapor shear stress

$$w_V = \frac{4 \times 1000}{3600 \times 2.897 \times \pi \times 50 \times 0.021^2} = 5.53 \text{ m/s}$$

$$G_{\text{tot}} = \frac{4 \times 2000}{3600 \times 50 \times \pi \times 0.021^2} = 32.08 \text{ kg/m}^2 \text{ s}$$

$$\alpha_\tau = 0.024 \times \frac{0.106}{0.021} \times \left(\frac{32.08 \times 0.021}{0.415 \times 10^{-3}}\right)^{0.8} \times 8.44^{0.43}$$

$$\times \frac{1 + \left(\frac{720}{2.897}\right)^{0.5}}{2} = 940 \text{ W/m}^2 \text{ K}$$

The higher value is valid $\alpha_g = 1463 \text{ W/m}^2 \text{ K}$.
The *Wallis factor J* is clearly lower than 1.5.

$$m = w_V \times \rho_V = 5.53 \times 2.897 = 16 \text{ kg/m}^2 \text{ s}$$

$$J = \frac{16}{9.81 \times 0.021 \times 2.897 \times (720 - 2.897)} = 0.77 < 1.5$$

Kern equation 2.1:

$$W = 1000 \text{ kg/h} = M_{\text{liq}}$$

$$G_{\text{Hi}} = \frac{1000}{0.5 \times 50 \times 3 \times 3600} = 0.0037 \text{ kg/m s}$$

$$\alpha_g = 0.925 \times 0.106 \times \left(\frac{720^2 \times 9.81}{0.415 \times 10^{-3} \times 3.7 \times 10^{-3}}\right)^{1/3} = 1461 \text{W/m}^2 \text{ K}$$

8.2.1.4 Condensation in vertical tubes
Equation 1.1:

$$M_{liq} = 1000 \text{ kg/h}$$

$$Re_{liq} = \frac{1000 \times 4}{3600 \times 50 \times \pi \times 0.021 \times 0.415 \times 10^{-3}} = 811.6$$

$$\alpha_g = 1.47 \times \frac{0.106}{31.35 \times 10^{-6}} \times \left(\frac{1}{811.6}\right)^{1/3} = 516 \text{ W/m}^2 \text{ K}$$

Calculation with equation 1.2 for $Re > 40$ for wavy condensate film:

$$\alpha_w = 1.007 \times \frac{0.106}{32.35 \times 10^{-6}} \times 811.6^{-2/9} = 744 \text{ W/m}^2 \text{ K}$$

Calculation according to equation 1.5 for shear stress

$$G_{av} = \sqrt{\frac{2000^2}{3}} \times \frac{4}{3600 \times 50 \times \pi \times 0.021^2} = 18.52 \text{ kg/m}^2 \text{ s}$$

$$\alpha_\tau = 0.0099 \times \frac{0.106}{0.021} \times \left(\frac{18.52 \times 0.021}{0.415 \times 10^{-3}}\right)^{0.9} \times 8.44^{0.5} \times \left(\frac{720}{2.897}\right)^{1/2}$$

$$\times \left(\frac{8.28 \times 10^{-6}}{0.415 \times 10^{-3}}\right)^{0.1} = 730 \text{ W/m}^2 \text{ K}$$

It is valid: $\alpha_w = 744 \text{ W/m}^2 \text{ K}$

Kern equation 2.1 for vertical condensation in tubes: $M_{liq} = 1000 \text{ kg/h}$
Calculation for laminar condensate film:

$$G_{Vi} = \frac{1000}{50 \times \pi \times 0.021 \times 3600} = 0.0842 \text{ kg/m s}$$

$$\alpha_g = 0.925 \times 0.106 \times \left(\frac{720^2 \times 9.81}{0.415 \times 10^{-3} \times 0.0842}\right)^{1/3} = 516 \text{ W/m}^2 \text{ K}$$

With turbulence correction factor f_c according to Equation 2.2 for turbulence

$$Re = \frac{4 \times 0.0842}{0.415 \times 10^{-3}} = 811.6 \quad \frac{Re}{4} = \frac{811.6}{4} = 202.9$$

$$f_c = 0.8 \times 202.9^{0.11} = 1.43$$

$$\alpha_t = 1.43 \times 516 = 740 \text{ W/m}^2 \text{ K}$$

Comparison of result for 2 t/h *vapors according to the Nitsche and the Kern method*

		Lumped	Zones
Horizontal at the tube bundle	Nitsche	$\alpha = 1330$ W/m^2 K	$\alpha = 1422$ W/m^2 K
	Kern	$\alpha = 1182$ W/m^2 K	$\alpha = 1205$ W/m^2 K
Vertical at the tube bundle	Nitsche	$\alpha = 774$ W/m^2 K	$\alpha = 815$ W/m^2 K
	Kern	$\alpha = 770$ W/m^2 K	$\alpha = 822$ W/m^2 K
Horizontal in the tubes	Nitsche	$\alpha = 1463$ W/m^2 K	$\alpha = 1564$ W/m^2 K
	Kern	$\alpha = 1461$ W/m^2 K	$\alpha = 1490$ W/m^2 K
Vertical in the tubes	Nitsche	$\alpha = 744$ W/m^2 K	$\alpha = 794$ W/m^2 K
	Kern	$\alpha = 740$ W/m^2 K	$\alpha = 788$ W/m^2 K

8.2.2 Parameters influencing the heat transfer coefficients

The heat transfer in film condensation is dependent on the condensate film thickness, the turbulence in the condensate film and the vapor shear stress.

In laminar region the heat transfer is only dependent on the heat conduction through the film thickness δ and the gravitation-driven heat transfer coefficient α_g is as follows:

$$\alpha_g = \frac{\lambda}{\delta} \left(\text{W/m}^2 \text{ K} \right)$$

In condensation in horizontal tubes and on a horizontal tube bundle there are thin condensate films and therefore the heat transfer coefficients are very high.

The film becomes thicker with increasing condensate load and the α-values fall.

In Figure 8.7 it is shown that the heat transfer coefficient in the individual zones of a condenser falls because the condensate film becomes thicker with increasing zone number.

In condensation in vertical tubes or on vertical surfaces the condensate film gets thicker downward.

This makes the heat transfer in laminar region worse but with increasing film thickness from Re > 40 waves built up on the condensate film and improves the heat transfer.

The heat transfer coefficient decreases much less with increasing condensate load.

From a Reynolds number Re = 1600 the turbulences in the condensate film are so large that the heat transfer coefficient increases.

With larger condenser vapor load or higher Reynolds numbers of the condensate the heat transfer in horizontal and vertical tubes is considerably improved by the vapor shear stress or the turbulent flow.

In Figure 8.8 the heat transfer coefficients of ethanol and cyclohexane as a function of the vapor rate is given.

A distinction is made between the shear stress-controlled and the gravitation-driven heat transfer.

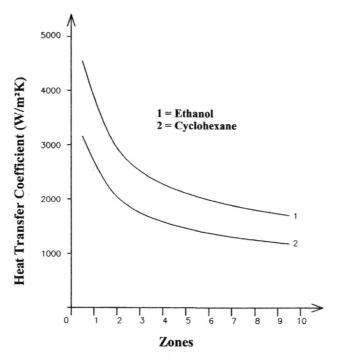

Figure 8.7 Heat transfer coefficients of ethanol and cyclohexane in the individual zones of a condenser.

In gravitation region, three different flow types are formed depending on the condensate Reynolds number.

Zone I up to $Re = 40$ with laminar condensate flow without waves

Zone II between $Re = 40$ and $Re = 1600$ with laminar wavy flow

Zone III from $Re = 1600$ with turbulent condensate flow

In zone I and II the heat transfer coefficient falls with increasing vapor rate because the condensate film becomes thicker. In section II the fall of the heat transfer coefficients is less than in section I because the heat transfer is improved by the waviness of the condensate film.

In zone III the heat transfer coefficient increases because the heat transfer resistance gets reduced by the turbulent condensate stream.

The curves 2 and 4 illustrate the improvement of the heat transfer by the vapor shear stress at larger vapor rates or higher flow velocities.

It also comes to an improvement of the heat transfer in condensation in horizontal tubes at high vapor loads by the vapor shear stress.

The condensation on the shell side of horizontal and vertical tube bundles is mostly gravitational driven. It comes to an improvement of the heat transfer by the vapor shear stress in the first rows only at a mass stream density of $m > 30 \text{ kg/m}^2 \text{ s}$.

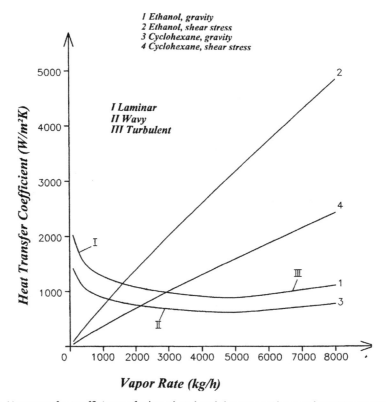

Figure 8.8 Heat transfer coefficients of ethanol and cyclohexane in the condensation in vertical tubes.

8.3 COMPARISON OF DIFFERENT CALCULATION MODELS

In a technical study, different calculation models for the determination of heat transfer coefficients at isothermal condensation for a certain water-cooled condenser were used.

The results are listed in Table 8.3.

Condenser dimensions:

Shell diameter 270 mm with 50 tubes 25 × 2, 6 m long, A = 23.5 m^2

Condensed material: Cyclohexane   Cooling: cooling water 25/35 °C

Heat resistance by fouling and heat conductivity through the wall: 0.00035.

In the table there is a distinction between the lumped and the zone wise calculation.

In the lumped calculation 50% of the total flow rate for the condensate load is input as average for the condensate load because the condensate rate of zero at the inlet rises up to 100% at the outlet.

An average condensate film thickness results from the average value between inlet and outlet, that is, at 50% of the total vapor rate.

Table 8.3 Heat transfer coefficients in isothermal film condensation

	Heat Transfer Coefficient Zone wise Calculation					Heat Transfer Coefficient Lumped Calculation				
Condensate kg/h	VDI W/m²K	Kern W/m²K	Colburn W/m²K	Huhn W/m²K	Nitsche W/m²K	VDI W/m²K	Kern W/m²K	Colburn W/m²K	Huhn W/m²K	Nitsche W/m²K
On horizontal bundle										
100	5050	4360			4908	4721	4077			4589
500	2953	2549			2870	2761	2384			2684
1000	2344	2023			2278	2191	1892			2130
2000	1860	1606			1808	1739	1502			1690
3000	1625	1403			1579	1519	1312			1477
4000	1476	1274			1435	1380	1192			1342
5000	1370	1183			1332	1281	1106			1245
In horizontal tubes										
100	6527	5391			5397	6418	5041			5047
500	3817	3153			3156	3753	2948			2951
1000	3029	2502			2505	2979	2340			2342
2000	1898	1986			1988	2364	1857			1859
3000	1733	1735			1737	1836	1622			1624
4000	1582	1576			1578	1624	1474			1475
5000	1657	1463			1465	1896	1368			1369
In vertical tubes										
100	1665	1510	24	1585	1574	1561	1412	46	1474	1414
500	1112	883	103	1045	1074	1035	825	199	986	1014
1000	1001	701	192	883	916	922	665	372	829	869
2000	993	556	359	805	789	905	520	695	730	745
3000	1061	505	517	798	863	958	454	1002	739	1006
4000	1185	513	670	806	1074	1035	412	1298	763	1304
5000	1338	544	819	819	1266	1422	383	1587	791	1594
On vertical bundles										
100	1747	1600		1661	1646	1639	1496		1540	1498
500	1143	936		1092	1118	1067	875		1030	1054
1000	989	743		918	953	922	694		866	904
2000	898	589		816	817	838	551		740	775
3000	874	515		798	731	820	481		731	708
4000	869	509		800	711	822	437		747	664
5000	873	509		809	720	833	406		769	543

In the zone wise approach the condenser is subdivided into 10 sections and for each zone the heat transfer coefficient for the condensate and vapor loading in this section is calculated and the required heat exchanger area for the zone is determined.

The heat transfer coefficients determined according to the lumped method are poorer in the gravitation-driven condensation because the average condensate film is thicker than that in the zone wise calculation. If however the shear stress determines the heat transfer, a too high value can result from the lumped method because of the greater vapor rate.

A zone wise calculation is recommended for the design.

The results of the calculated heat transfer coefficients with different models are shown in the following figures:

Figure 8.9: In condensation on horizontal tube bundle the results according to the four models are almost identical.

Figure 8.10: In condensation in horizontal tubes the results according to VDI-Wärmeatlas, Kern, and Nitsche only deviate marginally from another.

Figure 8.11: In condensation in vertical tubes the methods according to VDI and Nitsche are recommended since both consider the improvement by the shear stress. The models according to Kern and Huhn give too low heat transfer coefficients. The Colburn method should be avoided.

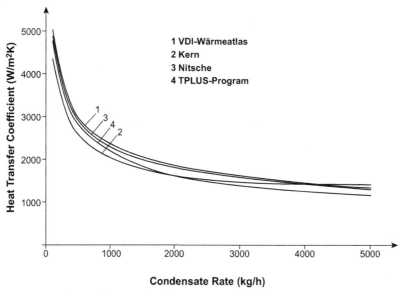

Figure 8.9 Heat transfer coefficients for the condensation on horizontal tube bundle. *(Literature sources: Colburn, Applied Mechanics Reviews, Vol. 5, 1952, Rev. 247; Huhn, Verfahrenstechnische Berechnungsmethoden Teil1, VCA Verlag, 1987; Kern, Process Heat Transfer, McGraw Hill, N.Y., 1950; VDI, VDI-Wärmeatlas, Ja1 – Ja19, VDI-Verlag, 1994.)*

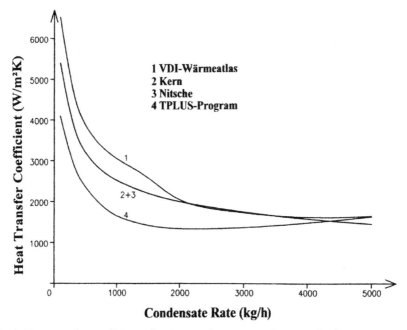

Figure 8.10 Heat transfer coefficients for the condensation in horizontal tubes. *(Literature sources: Colburn, Applied Mechanics Reviews, Vol. 5, 1952, Rev. 247; Huhn, Verfahrenstechnische Berechnungs-methoden Teil1, VCA Verlag, 1987; Kern, Process Heat Transfer, McGraw Hill, N.Y., 1950; VDI, VDI-Wärmeatlas, Ja1 – Ja19, VDI-Verlag, 1994.)*

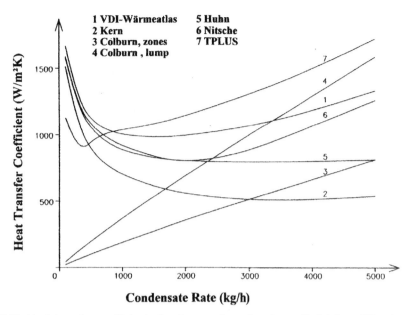

Figure 8.11 Heat transfer coefficients for the condensation in vertical tubes. *(Literature sources: Colburn, Applied Mechanics Reviews, Vol. 5, 1952, Rev. 247; Huhn, Verfahrenstechnische Berechnungsmethoden Teil1, VCA Verlag, 1987; Kern, Process Heat Transfer, McGraw Hill, N.Y., 1950; VDI, VDI-Wärmeatlas, Ja1 — Ja19, VDI-Verlag, 1994.)*

Figure 8.12: The calculation methods according to VDI, Huhn, and Nitsche give similar results.

Figure 8.13: In the calculation of the overall heat transfer coefficient U the different values of the heat transfer coefficients calculated according to different models are smoothed to a great extent by the heat transfer coefficient on the cooling water side and the fouling factors.

8.4 CONDENSATION OF VAPORS WITH INERT GAS [2]

Small amounts of noncondensable inert gases in the vapors to be condensed drastically reduce the heat transfer coefficient for the condensation of the vapors because the vapors to be condensed must diffuse to the cooled condensation surface by the noncondensable inert gas (see Figure 8.14).

The inert gas accumulates to the cooling wall because the vapors condense out so that an inert film builds up on the surface.

The vapors must diffuse through this gas film to the cold condensation surface.

This prevention in transportation deteriorates the heat transfer coefficient, for instance, for vapor in the presence of 1—10 Vol% air to 10—30% of the heat transfer coefficient of saturated vapor without inert gas.

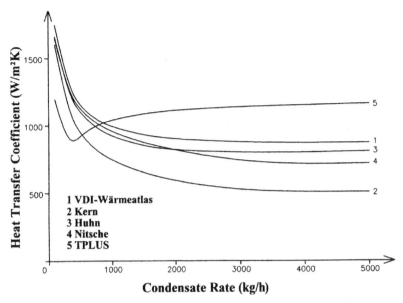

Figure 8.12 Heat transfer coefficients for the condensation in vertical tube bundle. *(Literature sources: Colburn, Applied Mechanics Reviews, Vol. 5, 1952, Rev. 247; Huhn, Verfahrenstechnische Berechnungs-methoden Teil1, VCA Verlag, 1987; Kern, Process Heat Transfer, McGraw Hill, N.Y., 1950; VDI, VDI-Wärmeatlas, Ja1 — Ja19, VDI-Verlag, 1994.)*

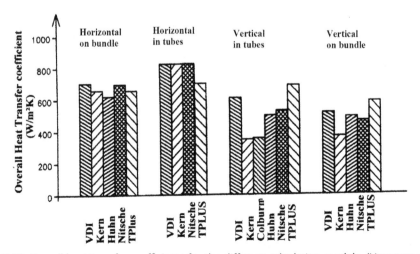

Figure 8.13 Overall heat transfer coefficients for the different calculation models. *(Literature sources: Colburn, Applied Mechanics Reviews, Vol. 5, 1952, Rev. 247; Huhn, Verfahrenstechnische Berechnungs-methoden Teil1, VCA Verlag, 1987; Kern, Process Heat Transfer, McGraw Hill, N.Y., 1950; VDI, VDI-Wärmeatlas, Ja1 — Ja19, VDI-Verlag, 1994.)*

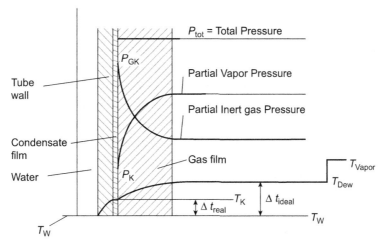

Figure 8.14 Condensation with inert gas.

In condensing out of the vapors the inert gas partial pressure toward the condensation surface increases and the vapor partial pressure decreases.

The reducing vapors—partial pressure results in the lowering of the dew or condensation temperature (see Figure 8.14).

Therefore, the effective temperature difference for the condensation is decreased and hence the condensation efficiency.

The inert gas functions as carrier for the vapors and deteriorates the efficiency in the liquidizing.

It must be considerably cooled deeper in order to liquefy the vapors. This is shown in Figure 8.15.

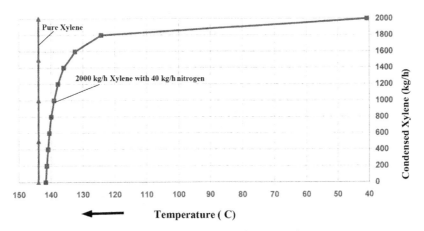

Figure 8.15 Condensed xylene rate as a function of the temperature.

Therefore, there are three effects that make the condensation efficiency worse in the presence of inert gases:

1. The transportation prevention by the gas film reduces the α-value for the condensation.
2. The reduction of the dew point reduces the effective Δt for the heat transfer.

$$\Delta t_{\text{real}} < \Delta t_{\text{ideal}}$$

Example for the reduction of dew point by inert gas
Total pressure: 950 mbar

Air content Vol%	H_2O—partial pressure mbar	Condensation temperature °C
0	950	98.2
5.2	900	96.71
10.5	850	95.15
15.8	800	93.51
21	750	91.78

3. It must be cooled considerably deeper in order to condense out the vapors (Figure 8.15).

In Figure 8.14 it is shown how the vapor partial pressure and the condensation temperature reduce toward the cooling wall and how the gas film builds up by the increase of the inert gas partial pressure.

The calculation method according to Colburn and Drew is very complex and is normally performed with a computer program.

An example for a hand calculation is given in Hewitt [3].

For hand calculations the following very simple calculation method for the determination of the heat transfer coefficient in the presence of inert gas can be adopted.

$$\alpha_{\text{corr}} = \frac{1}{\dfrac{Q_{\text{gas}}}{Q_{\text{tot}} \times \alpha_{\text{gas}}} + \dfrac{1}{\alpha_{\text{cond}}}} \ \text{W}/\text{m}^2\,\text{K}$$

Q_{gas} = convective heat duty for gas—or vapor cooling
Q_{tot} = total heat duty for the cooling and condensing
α_{gas} = heat transfer coefficient for the convective cooling
α_{cond} = heat transfer coefficient for the condensation

In addition to the condensation it must convectively cooled along the dew line and the heat transfer coefficient for the condensing is drastically reduced because of the low heat transfer coefficients for convective cooling of gases and vapors.

Example 4: Condensation of 2000 kg/h xylene vapors with 40 kg/h nitrogen

The condensation of pure xylene vapors occurs at the dew point of xylene at 143.9 °C.

The heat transfer coefficient is $\alpha_{\text{cond}} = 1680$ W/m² K and the overall heat transfer coefficient

$$U = 632 \ \text{W}/\text{m}^2\,\text{K}$$

In cooling with cooling water from 25 to 35 °C a driving temperature difference of 113.8 °C results for the isothermal condensation of xylene vapor.

Hence the required heat exchanger surface area A for $Q = 195$ kW is determined:

$$A = \frac{Q}{U \times \Delta t} = \frac{195,000}{632 \times 113.8} = 2.7 \text{ m}^2$$

For a mixture of 2000 kg/h xylene and 40 kg/h nitrogen the following calculation results: The mixture must be cooled down to 40 °C in order to liquefy the 2000 kg/h xylene. This reduces the effective temperature difference from 113.8 to 71.7 °C.

Calculation of the corrected heat transfer coefficient: α_{corr}

$$Q_{gas}/Q_{tot} = 0.316 \qquad \alpha_{gas} = 195 \text{ W/m}^2 \text{ K} \qquad \alpha_{cond} = 1680 \text{ W/m}^2 \text{ K}$$

$$\alpha_{corr} = \frac{1}{\frac{0.316}{195} + \frac{1}{1680}} = 451 \text{ W/m}^2 \text{ K}$$

Calculation of the overall heat transfer coefficient U with $\alpha_{Water} = 2956$ W/m^2 K

$$\frac{1}{U} = \frac{1}{2956} + \frac{1}{451} + 0.00034 \quad U = 345 \text{ W/m}^2 \text{ K}$$

Required heat exchanger surface area A for $Q = 312$ kW:

$$A = \frac{312,000}{71.7 \times 345} = 12.6 \text{ m}^2$$

8.5 CONDENSATION OF MULTICOMPONENT MIXTURES (SEE CHAPTER 7)

For the design of condensers for multicomponent mixtures the condensation curve between the dew and bubble point is needed. The condensation function is not isotherm.

The calculation of such condensation curves of multicomponent mixtures is not simple, especially for nonideal mixtures.

The condensation curve for the binary mixture benzene—toluene is shown in Figure 8.16.

The vapors enter the condenser with $y_1 = 30$ mol% benzene. The dew point is at 104 °C.

The out-condensed liquid is enriched with the high-boiling component toluene and the composition of the lighter component benzene drops from y_1 in the vapor to x_1 in the condensate. Due to the preferred out-condensation of the high-boiling toluene the benzene composition of the vapors rises to y_2 with the equilibrium composition x_2 in the condensate.

After cooling to the boiling point at 98 °C the vapor composition y_1 and the exiting condensate have the same composition as the vapors at the inlet: $y_1 = x_3$.

The temperature drop from dew point to bubble point has an influence on the effective temperature difference for the heat transfer.

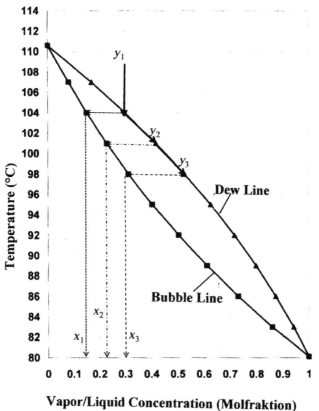

Figure 8.16 Bubble and dew line for the mixture of benzene–toluene.

In nonlinear shape of the temperature in the temperature heat load diagram the average weighted temperature difference, WMTD, must be determined (see Example 8).

In comparison with the isothermal condensation of individual components the heat transfer coefficients in condensing of mixtures are clearly worse because the light-boiling components on the cooling surface prevent the condensing of the higher boiling components or because the mixture must be convectively cooled from the dew point to the bubble point.

The deterioration of the heat transfer coefficients is shown in the following overview.

Components	Benzene	Benzene + toluene	Benzene + toluene + xylene
Heat transfer coefficient (W/m² K)	1983	1662	1410
Components	Butane	Butane + toluene	Butane + toluene + xylene
Heat transfer coefficient (W/m² K)	2211	1405	1218

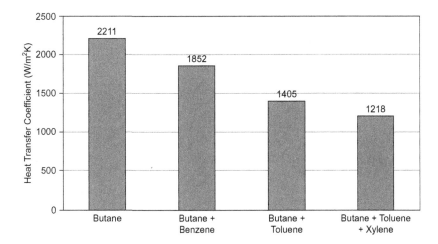

The heat transfer coefficient for the convective cooling of gases or vapors is much worse than the α-value for the condensing.

Therefore the corrected α-value considering the convective cooling and condensation is less than the heat transfer coefficient for pure condensing.

The deterioration of the heat transfer coefficient by the convective cooling is dependent on the heat load Q_{gas} for the gas or vapor cooling.

An improvement of the convective heat transfer coefficient for the vapor cooling is achieved by the higher flow velocity of the vapors.

In the following it is shown how to estimate the correction α-value for the condensation with a simple method if the ratio Q_{gas}/Q_{tot} is known.

In Figure 8.17 the reduction of the heat transfer coefficient with increasing gas cooling is depicted.

$$\alpha_{corr} = \frac{\alpha_{cond}}{1 + \frac{Q_{gas} \times \alpha_{cond}}{Q_{tot} \times \alpha_{gas}}} = \frac{1}{\frac{Q_{gas}}{Q_{tot} \times \alpha_{gas}} + \frac{1}{\alpha_{cond}}} \; (W/m^2\,K)$$

Q_{gas} = convective heat duty for the gas and vapor cooling
Q_{tot} = total heat duty for the cooling and condensation
α_{gas} = heat transfer coefficient for the convective cooling
α_{cond} = heat transfer coefficient for the condensation

Example 5: Condensation of a multicomponent mixture considering the gas cooling

$\alpha_{gas} = 50\ W/m^2\,K$ $\qquad$ $\alpha_{cond} = 2500\ W/m^2\,K$ $\qquad$ $Q_{gas}/Q_{tot} = 0.05$

$$\alpha_{corr} = \frac{1}{\frac{0.05}{50} + \frac{1}{2500}} = 714\ W/m^2\,K \quad \alpha_{corr} = \frac{2500}{1 + \frac{0.05 \times 2500}{50}} = 714\ W/m^2\,K$$

By the vapor cooling the high α-value of 2500 W/m^2 K for the condensation is reduced to 714 W/m^2 K.

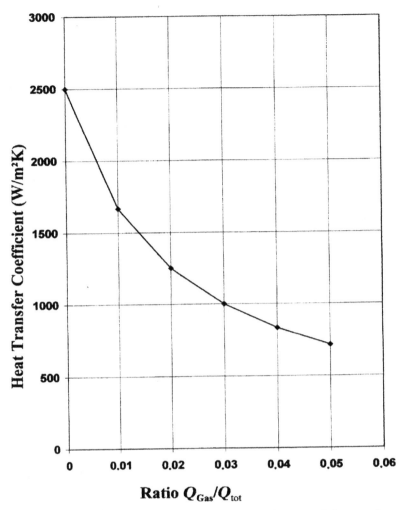

Figure 8.17 Heat transfer coefficients for the condensation as function of the partial gas cooling.

Example 6: Condensation of a mixture of 1000 kg/h benzene and 1000 kg/h xylene

Dew point: 121.8 °C   Bubble point: 96.1 °C

$Q_{gas} = 25.5$ kW $Q_{tot} = 238$ kW $Q_{gas}/Q_{tot} = 0.109$

$\alpha_{cond} = 2000$ W/m² K $\alpha_{gas} = 238$ W/m² K

$$\alpha_{corr} = \frac{1}{\frac{0.109}{238} + \frac{1}{2000}} = 1041 \text{ W/m}^2 \text{ K}$$

The convective cooling from the dew point 121.8 °C to the bubble point of 96.1 °C reduces the heat transfer coefficient from 2000 to 1041 W/m² K.

Example 7: Condensation of a hydrocarbon mixture containing inert gas

Nitrogen	50 kg/h	Inlet: 132 °C
Pentane	1000 kg/h	Outlet: 30 °C
Octane	2000 kg/h	Condensed: 5968 kg/h liquid
Nonane	3000 kg/h	Gas outlet: 82 kg/h
Total:	6050 kg/h	

$Q_{gas} = 350.2$ kW $Q_{tot} = 881.2$ kW $Q_{gas}/Q_{tot} = 0.397$ $A = 66.6$ m^2

$\alpha_{gas} = 260$ W/m^2 K $\alpha_{cond} = 1500$ W/m^2 K   WMTD $= 45.26$ °C

$$\alpha_{corr} = \frac{1}{\frac{0.397}{260} + \frac{1}{1500}} = 455.5 \text{ W/m}^2 \text{ K}$$

Cooling water: $\alpha = 2690$ W/m^2 K

$$\frac{1}{U} = \frac{1}{2690} + \frac{1}{455.5} + 0.00034 \quad U = 344 \text{ W/m}^2 \text{ K}$$

$$A = \frac{Q_{tot}}{U \times \text{WMTD}} = \frac{881,200}{344 \times 42.26} = 60.6 \text{ m}^2 \approx 10\% \text{ reserve}$$

Example 8: Condensation of a mixture of 500 kg/h pentane and 4000 kg/h decane

Dew point: 167.6 °C   Bubble point: 96.1 °C   Heat duty $Q_{gas} = 588$ kW

Cooling water: $\alpha = 4900$ W/m^2 K.

$\alpha_{cond} = 1700$ W/m^2 K $\alpha_{gas} = 370$ W/m^2 K $Q_{gas}/Q_{tot} = 0.366$ $A = 14$ m^2

Calculation of the corrected heat transfer coefficient:

$$\alpha_{corr} = \frac{1}{\frac{0.366}{370} + \frac{1}{1700}} = 634 \text{ W/m}^2 \text{ K}$$

The heat transfer coefficient deteriorates from 1700 to 634 W/m^2 K by the convective cooling of the vapors from dew point to bubble point.

Calculation of the overall heat transfer coefficient:

$$\frac{1}{U} = \frac{1}{4900} + \frac{1}{634} + 0.00034 \quad U = 471 \text{ W/m}^2 \text{ K}$$

Calculation of the average weighted temperature difference, WMTD:

Due to the nonlinear shape of the temperature over the heat duty the weighted average temperature must be determined (see Chapter 2).

Hence the temperature–heat duty curve shown in Figure 8.18 is then subdivided into four zones with an approximate linear shape.

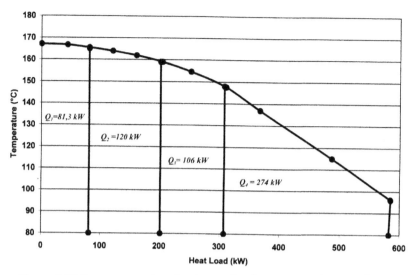

Figure 8.18 Temperature–heat load curve for the pentane–decane mixture.

For each zone the heat duty and the corrected temperature difference, CMTD, is determined and then the average weighted temperature difference is calculated.

Calculation of the weighted temperature difference, WMTD:

$Q_1 = 81.3$ kW $\mathrm{CMTD}_1 = 136.5\,°C$ $Q_1/\mathrm{CMTD}_1 = 0.595$
$Q_2 = 120$ kW $\mathrm{CMTD}_2 = 132.2\,°C$ $Q_2/\mathrm{CMTD}_2 = 0.908$
$Q_3 = 106$ kW $\mathrm{CMTD}_3 = 123.1\,°C$ $Q_3/\mathrm{CMTD}_3 = 0.86$
$Q_4 = 274$ kW $\mathrm{CMTD}_4 = 89.35\,°C$ $Q_4/\mathrm{CMTD}_4 = 3.066$
$\ni Q = 588$ kW $\ni Q/\mathrm{CMTD} = 5.43$

$$\mathrm{WMTD} = \frac{\sum Q}{\sum Q/\mathrm{CMTD}} = \frac{588}{5.43} = 108.3\,°C$$

Under the assumption of a linear temperature curve a CMTD = 97.5 °C results. This value is wrong!

Calculation of the required area:

$$A = \frac{588,000}{471 \times 108.3} = 11.5\ \mathrm{m}^2$$

Differential and integral condensation of multicomponent mixtures

The condensation on horizontal tube bundles is not an equilibrium or integral condensation in which vapor and liquid are in equilibrium. The high-boiling components that are condensed out at the inlet of the condenser exit from the tubes at the bottom of the heat exchanger and no longer participate in equilibrium.

This differential or nonequilibrium condensation is subjected to worse condensation conditions.

The light-boiling components build up in the condensation process and reduce the dew point.

The effective temperature difference becomes smaller and it must be cooled deeper.

This is highly predominant in condensation of wide boiling components, for instance, ammonia—water.

In Figure 8.19 an equilibrium condenser with condensation in vertical tubes and a differential condenser with stage condensation on horizontal tube bundle is shown.

In Figure 8.20 the condensation process for a mixture of the light hydrocarbons propane, butane, and pentane in air in integral and stage condensation is depicted.

In the stage condensation it must be cooled much deeper in order to liquefy the gasoline vapor.

8.6 MISCELLANEOUS

In the following, some problems in the selection and design of condensers are treated.

8.6.1 Influence of pressure loss in the condenser

Dew and bubble point of multicomponent mixtures are determined by the composition of the vapors and the pressure.

Due to the pressure loss in the condenser the pressure in the system drops.

Therefore the dew and bubble point are lowered.

This reduces the effective temperature difference for the condensation.

Example 9: Condensation of 2 t/h benzene and 2 t/h toluene with cooling water 25/35 °C at different pressures

Pressure = 1 bar:	Dew point = 97.48 °C	Bubble point = 90.94 °C	CMTD = 64.02 °C
Pressure = 0.95 bar:	Dew point = 95.68 °C	Bubble point = 89.12 °C	CMTD = 62.2 °C
Pressure = 0.5 bar:	Dew point = 74.56 °C	Bubble point = 67.76 °C	CMTD = 40.9 °C

8.6.2 Condensate drain pipe

The exit stream from the condenser to the condensate accumulator should be self-aerated.

This is achieved with a Froude number $Fr < 0.31$.

$$Fr = \frac{w}{\sqrt{g \times d}} = 2.4 \times \left(\frac{H}{d}\right)^{1.5} < 0.31$$

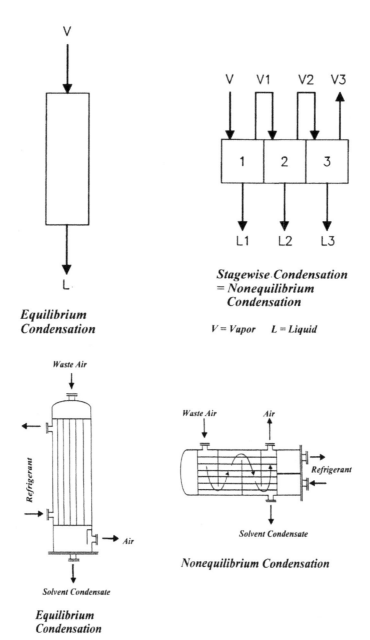

Figure 8.19 Integral condenser with equilibrium condensation and differential condenser with stagewise nonequilibrium condensation.

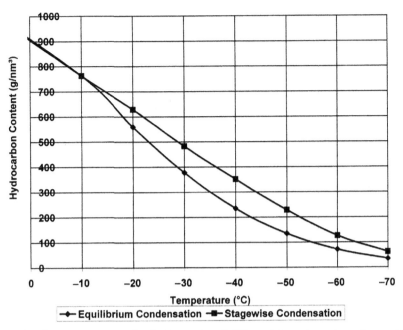

Figure 8.20 Condensation process for the integral and differential condensation of a gasoline vapor mixture in an air–hydrocarbon mixture.

The required nozzle diameter d for a condensate exit rate V (m³/h) is determined as follows:

$$d = 0.0422 \times V^{0.4} (m)$$

The ratio liquid height H over the nozzle to the nozzle diameter is thereby $H/d = 0.25$.

In order to avoid after evaporation by the static negative pressure in the exit pipe a liquid height H_{req} over the exit nozzle is necessary.

$$H_{req} = \frac{2 \times w^2}{g}$$

$w = $ liquid velocity (m/s)

Example 10: Required condensate exit pipe diameter for $V = 4$ m³/h condensate

$d = 0.0422 \times 4^{0.4} = 0.0735 \text{ m} = 73.5 \text{ mm}$ $H = 0.25 \times 0.0735 = 0.018 \text{ m} = 18.4 \text{ mm}$

$$w = \frac{4/3600}{0.0735^2 \times (\pi/4)} = 0.263 \text{m/s}$$

$$Fr = \frac{0.263}{\sqrt{9.81 \times 0.0735}} = 0.31 = 2.4 \times \left(\frac{18.4}{73.5}\right)^{1.5} = 0.3$$

$$H_{req} = 2 \times \frac{0.263^2}{9.81} = 0.014 \text{ m} = 14 \text{ mm}$$

8.6.3 Maximal vapor flow capacity

In condensation in vertical or horizontal tubes the maximal adiabatic mass flux G must be examined at the inlet from a large vapor pipe into the several small pipes of the condenser. The adiabatic flow capacity is calculated with the adiabatic exit flow equation for nozzles.

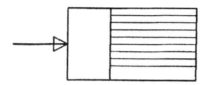

$$G = \varepsilon \times F \times \psi \times \sqrt{2 \times P_1 \times \rho_1} \, (\text{kg/s})$$

$$\psi = \sqrt{\frac{\kappa}{\kappa - 1} \times \left[\left(\frac{P_2}{P_1} \right)^{2/\kappa} - \left(\frac{P_2}{P_1} \right)^{\frac{\kappa+1}{\kappa}} \right]}$$

$F =$ flow cross-section (m^2)
$G =$ adiabatic flow capacity (kg/s)
$P_1 =$ pressure before the inlet into the pipes (Pa)
$P_2 =$ pressure at the outlet from the tubes (Pa)
$\varepsilon =$ flow contraction $(0.6-0.8)$
$\kappa =$ adiabatic exponent
$\psi =$ adiabatic flow factor

The condensation efficiency is limited by the adiabatic flow capacity especially in condensation in vacuum.

Example 11: Adiabatic flow capacity of a condenser DN 700 with 364 tubes 25 × 2

Vapor mole weight $M = 200$ Gas density $\rho_1 = 0.129$ kg/m^3 $\kappa = 1.4$
$P_1 = 2000$ Pa $P_2 = 1500$ Pa $P_2/P_1 = 0.75$
$\psi = 0.4279$ $\varepsilon = 0.8$

Calculation of the adiabatic flow capacity:

$$G = 0.8 \times 0.000346 \times 0.4279 \times \sqrt{2 \times 2000 \times 0.129} = 0.00269 \text{ kg/s pro Rohr} = 9.68 \text{ kg/h}$$

Maximum total throughput in 364 tubes: 3525 kg/h
Calculation of the condensation performance M (kg/h):
$A = 114.3$ m^2 area in condenser ($L = 4$ m)

$U = 600$ W/m^2 K   LMTD $= 25\,°$C   Latent heat $= 150$ Wh/kg

$$Q = U \times A \times \text{LMTD} = 600 \times 114.3 \times 25 = 1714,500 \text{ W}$$

$$M = \frac{Q}{r} = \frac{1714,500}{150} = 11,430 \text{ kg/h}$$

Thermally 11,430 kg/h can be condensed, but through the adiabatic flow capacity the throughput is limited to 3525 kg/h.

___

8.6.4 Top condensers on columns [11]

The arrangement of condensers at the top of distillation columns has some advantages (Figure 8.21):

The pumping of the reflux to the column top is not necessary.

No pumping is required and no pipe lines and no steel supports for the condenser and the accumulator.

No energy is required for the back pumping of the reflux.

Due to the elimination of the long vapor pipes the pressure loss is reduced.

This is a great advantage especially in vacuum distillations.

The low holdup in the condenser simplifies the separation in discontinuous batch distillation.

Most important for the proper function of a top condenser is the correct hydraulic design of the reflux line. It is a process of gravity-driven boiling liquid from the condenser to the column.

The driving height H must be greater than the pressure loss between the points A and B.

H is the difference in height between the condenser outlet nozzle of the condenser and the inlet nozzle of the column. The vapor density ρ_V is neglected.

$$H \times \rho_{\text{liq}} \times g = \Delta P_{\text{cond}} + \Delta P_{\text{RL}} + \Delta P_{\text{M}} + \Delta P_{\text{CV}}$$

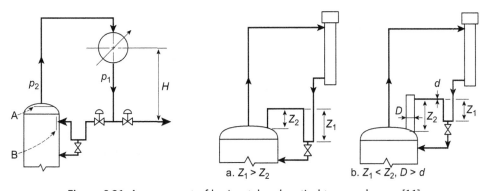

Figure 8.21 Arrangement of horizontal and vertical top condensers [11].

ΔP_{cond} = pressure loss in the condenser (Pa)
ΔP_{RL} = pressure loss in the reflux line (Pa)
ΔP_M = pressure loss of a flow measuring device (Pa)
ΔP_{CV} = pressure loss of a control valve (Pa)

In order to avoid large additional heights for the increase of the difference in height H the pressure losses in the reflux line is minimized. Flow measuring devices with low pressure loss are used and only in special cases a control valve is installed. The reflux line is dimensioned for a self-venting flow.

Example 12: Design of a reflux line for $V = 28.26 \, m^3/h$

Calculation of the diameter for a self-venting flow:

$$d = 0.0422 \times 28.26^{0.4} = 0.160 \, m = 160 \, mm$$

Calculation of the pressure loss for the reflux:
Flow velocity $w = 0.39 \, m/s$.

Pipe length $L = 10 \, m$ Liquid density $\rho = 800 \, kg/m^3$

$$\Delta P_R = \left(0.04 \times \frac{10}{0.16}\right) \times \frac{0.39^2 \times 800}{2} = 152 \, Pa$$

Pressure loss in the condenser and in the flow measuring device:

$\Delta P_{cond} = 200 \, Pa$ $\Delta P_M = 6000 \, Pa$
Total pressure loss $\Rightarrow \Delta P = 200 + 152 + 6000 = 6352 \, Pa$

Required height difference H_{req}:

$$H_{req} = \frac{\sum \Delta P}{g \times \rho_{liq}} = \frac{6352}{9.81 \times 800} = 0.81 \, m \qquad \text{Chosen: } H = 1.5 \, m$$

Due to load variations in the column disturbances in the hydraulic equilibrium and the reflux variations can occur. This reduces the efficiency of the fractionation.

The reflux variations can be prevented by installing a control valve in the reflux line, however sufficient height H for the additional pressure loss in the control valve must be provided.

Example 13: Design of a reflux line with control valve

Data of Example 12
Additional pressure loss in the control valve $\Delta P_{CV} = 0.2 \, bar = 20,000 \, Pa$
$\Rightarrow \Delta P = \Delta P_{cond} + \Delta P_{RL} + \Delta P_M + \Delta P_{CV} = 200 + 152 + 6000 + 20,000 = 26,352 \, Pa$

$$H_{req} = \frac{26,352}{9.81 \times 800} = 3.36 \, m$$

In order to reduce the required height H a flow measuring device and a control valve of low pressure loss is used.

$$\Delta P_M = 1000 \text{ Pa} \qquad\qquad \Delta P_{RV} = 5000 \text{ Pa}$$
$$\ni \Delta P = \Delta P_{Kon} + \Delta P_R + \Delta P_M + \Delta P_{RV} = 200 + 152 + 2000 + 6000 = 8352 \text{ Pa}$$

$$H_{req} = \frac{8352}{9.81 \times 800} = 1.06 \text{ m} \qquad \text{Chosen: } H = 1.5 \text{ m}$$

The chosen driving height H should always be substantially higher than the calculated required height H_{req}.

A locking siphon is installed at the reflux inflow in the column in order to prevent vapor short circuit. The height Z_1 must be greater than Z_2 or the diameter D of Z_2 must be greater than the diameter of the siphon in order to prevent the canceling out of the siphon.

This lock can cause problems if after condensing two liquid phases with severely different densities form.

Siphon in flow disturbances occurs if the two siphon arms are filled with liquids of different weights. This is why draining facilities must be provided.

8.6.5 Calculation of the pressure loss ΔP_T in condensing in the tubes [4]

The friction pressure loss ΔP_T is calculated as follows:

$$\Delta P_T = \frac{w_m^2 \times \rho}{2} \times \left(f \times \frac{L}{d} + 1.5 \right) (\text{Pa})$$

$$w_m = \frac{w_{in} + w_{out}}{2} (\text{m/s})$$

d = tube diameter (m) $\qquad\qquad$ f = friction factor $\qquad$ L = tube length (m)
w_m = average flow velocity (m/s) $\qquad\qquad\qquad\qquad$ w_{in} = inlet velocity (m/s)
w_{out} = outlet velocity (m/s) $\qquad\qquad\qquad\qquad\qquad$ ρ_V = gas density (kg/m³)

Due to the condensation of the vapors a pressure regain is achieved by the negative acceleration.

The vapor velocity is reduced to 0 of the inlet stream velocity.

The pressure regain by the negative acceleration is calculated as follows:

$$\Delta P_{acc} = m \times (w_{out} - w_{in})$$

$$m = \frac{M}{3600 \times d^2 \times \pi/4 \times n} = w_{in} \times \rho (\text{kg/m}^2 \text{ s})$$

M = vapor inlet rate (kg/h)
m = mass stream density (kg/m² s)
n = number of tubes

Example 14: Pressure loss calculation in a condenser

M = vapor inlet rate to condenser = 4 t/h

Horizontal condenser with 72 tubes $a_{Tube} = 0.0249 \text{ m}^2$
25 × 2. 4 m long
$\rho_{liq} = 720 \text{ kg/m}^3$ $\rho_V = 1.623 \text{ kg/m}^3$ Friction factor $f = 0.04$

G_V = vapor rate at the inlet = 4000/1.623 = 2464.5 m^3/h

$$w_{in} = \frac{G_V}{3600 \times a_{Tube}} = \frac{2464.5}{3600 \times 0.0249} = 27.5 \text{ m/s} \quad w_{out} = 0 \text{ m/s}$$

$$w_m = \frac{27.5 + 0}{2} = 13.75 \text{ m/s}$$

Calculation of the friction pressure loss ΔP_T:

$$\Delta P_R = \frac{13.75^2 \times 1.623}{2} \times \left(0.04 \times \frac{4}{0.021} + 1.5\right) = 1399 \text{ Pa} = 14 \text{ mbar}$$

Calculation of the pressure regain ΔP_{acc}:

$$m = 27.5 \times 1.623 = 44.6 \text{ kg/m}^2 \text{ s} \quad \Delta P_{acc} = 44.6 \times (0 - 27.5) = 1226 \text{ Pa} = 12.3 \text{ mbar}$$

Effective pressure loss $\Delta P_{eff} = \Delta P_T - \Delta P_{acc} = 14 - 12.3 = 1.7 \text{ mbar}$.

8.6.6 Maximum condensate load in horizontal tubes [7]

If the condensate load is too high in horizontal tubes the condenser becomes flooded.

In order to avoid this, the condensate height h in the tube at the outlet should not exceed a fourth of the tube diameter di.

The allowable condensate load W_{all} for the condition $d_i = 0.25$ is determined as follows:

$$W_{all} = 1570 \times n \times \rho \times d_i^{2.56} (\text{kg/h})$$

n = number of tubes
ρ = condensate density (kg/m^3)

Example 15: Allowable condensate load in horizontal tubes

Condenser with 186 tubes 25 × 2

$d_i = 21 \text{ mm}$ $L = 4 \text{ m}$ $A = 58 \text{ m}^2$ $\rho = 614.3 \text{ kg/m}^3$
$W_{all} = 1570 \times 186 \times 614.3 \times 0.021^{2.56} = 9092 \text{ kg/h} \quad m = 39.2 \text{ kg/m}^2 \text{ s}$
Condenser with 80 tubes 57 × 3.5
$d_i = 50 \text{ mm}$ $L = 4 \text{ m},$ $A = 57.3 \text{ m}^2$ $\rho = 614.3 \text{ kg/m}^3$
$W_{all} = 1570 \times 80 \times 614.3 \times 0.05^{2.56} = 36,036 \text{ kg/h} \quad m = 408 \text{ kg/m}^2 \text{ s}$

By the selection of larger pipe diameters the capacity in condensing in horizontal tubes can be increased.

8.6.7 Flood loading in reflux condensers [4,12−14]

In the vertical reflux condensers installed on stirred tanks for the boiling in the reflux the allowable flood loading must be checked alongside with the thermal design. In these apparatus the vapors flow upward and the condensate runs downward in the opposite direction to the vapors.

Congestion of the condensate occurs if the vapor velocity is too high.

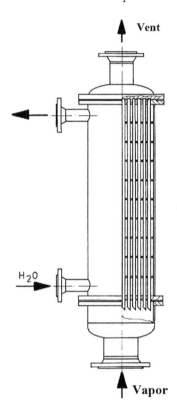

The allowable vapor load G is determined according to English and others as follows:

$$G = 246 \times \frac{d_i^{0.3} \times \rho_{\text{liq}}^{0.46} \times \sigma^{0.09} \times \rho_V^{0.5}}{\eta_{\text{liq}}^{0.14} \times \cos \delta^{0.32} \times (C/V)^{0.07}} \left(\text{kg}/\text{m}^2 \text{ h}\right)$$

$$G_{\text{cond}} = G \times A_{\text{tube}} (\text{kg}/\text{h})$$

The following is valid for the allowable vapor load of the reflux nozzle on the stirred tank:

$$G_{\text{nozz}} = G \times \frac{d_{\text{noz}}^{0.3}}{d_i^{0.3}} \left(\text{kg}/\text{m}^2 \text{ h}\right)$$

d_i = inner tube diameter (mm)

d_{nozz} = reflux nozzle diameter (mm)

V = vapor rate (kg/h)

G = allowable vapor load for the inner tubes (kg/m^2 h)

G_{cond} = allowable vapor load of the condenser (kg/h)

G_{nozz} = allowable nozzle load (kg/m^2 h)

C = condensate rate (kg/h)

ρ_V = vapor density (kg/m^3)

ρ_{liq} = liquid density (kg/m^3)

η_{liq} = liquid viscosity (mPas)

σ = surface tension (N/m)

δ = tube chamfered angle ($^\circ$)

Notice: The allowable flood load can be increased if the inner tubes of the condenser are pushed through the tube sheet and chamfered. The same holds for the reflux nozzle.

Example 16: Calculation of the allowable vapor load for pushed through and chamfered tubes

Reflux condenser with 110 tubes 20 × 2, d_i = 16 mm, chamfered angle δ = 75°

$$\rho_{liq} = 958 \text{ kg/m}^3 \qquad \rho_V = 0.58 \text{ kg/m}^3 \qquad \sigma = 5.8 \text{ N/m} \qquad \eta_{liq} = 0.28 \text{ mPas} \qquad C/V = 1$$

$$G = 246 \times \frac{16^{0.3} \times 958^{0.46} \times 5.8^{0.09} \times 0.58^{0.5}}{0.28^{0.14} \times 0.259^{0.32} \times 1^{0.07}} = 21835 \text{ kg/m}^2\text{h}$$

$$\text{Rohrquerschnitt } F = 110 \times 0.016^2 \times 0.785 = 0.022 \text{ m}^2$$

$$G_K = G \times F = 21,835 \times 0.022 = 480 \text{ kg/h}$$

Allowable nozzle load for a chamfered nozzle with d_{nozz} = 150 mm

$$G_{nozz} = 21,835 \times \frac{150^{0.3}}{16^{0.3}} = 42,732 \text{ kg/m}^2\text{h}$$

$$A_{nozz} = 0.15^2 \times 0.785 = 0.0177 \text{ m}^2$$

$$G_{d=150} = 42,732 \times 0.0177 = 756 \text{ kg/h}$$

Without chamfered tubes and nozzles the following allowable loadings result:

Chamfered angle $\delta = 0$	$\cos \delta = 1$
$G = 14,171 \text{ kg/m}^2 \text{ h}$	$G_{cond} = 14,171 \times 0.022 = 312 \text{ kg/h}$
$G_{nozz} = 27,734 \text{ kg/m}^2 \text{ h for } d = 150 \text{ mm}$	$G_{d=150} = 28,177 \times 0.0177 = 491 \text{ kg/h}$

NOMENCLATURE

c_p specific heat capacity (Wh/kg K)

d_o tube outer diameter (m)

d_i inner tube diameter (m)

G condensate rate per unit periphery (kg/m s)

g gravitational acceleration (m/s^2)

G_m average mass stream density (kg/m^2 s)

L characteristic length

l_{tube} tube length (m)

M_{liq} condensate mass stream (kg/h)

M_{tot} total mass stream (kg/h)

n number of tubes

Pr Prandtl number

Re_{liq} condensate Reynolds number

W vapor rate (kg/h)

w_V vapor velocity (m/s)

α_g gravity-controlled heat transfer coefficient (W/m^2 K)

α_t heat transfer coefficient for turbulent condensate flow (W/m^2 K)

α_w heat transfer coefficient for wavy flow (W/m^2 K)

α_τ shear stress-controlled heat transfer coefficient (W/m^2 K)

η_{liq} dynamic viscosity of the liquid phase (Pas)

η_V dynamic viscosity of the vapor phase (Pas)

λ condensate heat conductivity (W/m K)

ρ_{liq} density of the liquid phase (kg/m^3)

ρ_V density of the vapor phase (kg/m^3)

ν_{liq} kinematic viscosity of the liquid phase (m^2/s)

REFERENCES

[1]   D.Q. Kern, Process Heat Transfer, McGraw-Hill, New York, 1950.

[2]   W.M. Rohsenow, J.P. Hartnett, Handbook of Heat Transfer, McGraw-Hill, New York, 1973.

[3]   G.F. Hewitt, G.L. Shires, T.R. Bott, Process Heat Transfer, CRC Press Boca Raton, 1994.

[4]   N.P. Chopey, Heat Transfer, in: Handbook of Chemical Engineering Calculations, McGraw-Hill, New York, 1993.

[5]   W. Nusselt, Ver. Deutsch. Ing. 60 (1916) 541.

[6]   G.H. Gilmore, Chem. Eng. (April 1953).

[7]   R.C. Lord, P.E. Minton, R.P. Slusser, Design parameters for condensers and reboilers, Chem. Eng. 23 (3) (1970) 127.

[8]   A. Devore, How to design multitube condensers, Pet. Ref. 38 (June 1959) 205.

[9]   M. Hadley, Kondensation binärer Dampfgemische unter dem Einfluß der vollturbulenten ömung, in: Fortschr.Ber.VDI Reihe, 3, VDI-Verlag Düsseldorf, 1997. Nr. 468.

[10] A.A. Durand, C.A.A. Guerrero, E.A. Ronces, Gravity effects in horizontal condensers, Chem. Eng. 109 (April 2002) 91—94.

[11] R. Kern, How to design overhead condensing systems, Chem. Eng. (September 1975).

[12] K.G. English, u. andere, Chem. Eng. Progr. 59 (1963) 7.

[13] K. Feind, Verfahrenstechnik 5 (1971) 10.

[14] F.A. Zenz, R.E. Lavin, Hydrocarbon Process. 44 (3) (1965).

CHAPTER 9

Design of Evaporators

Contents

Heat Exchanger Design Guide
http://dx.doi.org/10.1016/B978-0-12-803764-5.00009-2

© 2016 Elsevier Inc.
All rights reserved.

9.1 EVAPORATION PROCESS

In evaporation, depending on the difference between the hot wall and the boiling temperature, the physical properties and the stream conditions, three different forms of boiling can occur: *pool boiling*, *flow boiling*, and *flash evaporation*.

9.1.1 Pool boiling

The evaporation of a stationary liquid in a vessel that is heated from the bottom is called pool boiling.

If the temperature differences between the heating wall T_W and the boiling medium T_B are low, the evaporation takes place from the surface area. For this "surface flash evaporation," the heat transfer coefficients are determined with the relationships for free convection (Figure 9.8).

The heat transfer coefficients are small in comparison to nucleate boiling.

At higher temperature differences, $T_W - T_B$, bubbles build up at the heating surface, which break up at a certain size and flow upward. By further increasing of the temperature difference or the heat flux density, larger turbulence occurs in the liquid and the heat transfer increases. This situation is called nucleate boiling.

In technical apparatus, work is done in pool boiling at temperature differences $T_W - T_B$ of 10–30 °C. The heat transfer coefficients lie in the range of 1000–3000 W/m² K for organic media and at 6000–12,000 W/m² K for water at a pressure of 1 bar (Figure 9.9).

In a too high temperature difference between the wall and the boiling medium, a coherent layer of vapor is build up on the heating surface. Due to the poor thermal heat conductivities of the vapors compared with liquids, the heat transfer coefficients become severely deteriorate.

This mechanism is called film evaporation (Figure 9.8).

In unstable film evaporation, the vapor film is broken by the liquid in certain points.

The stable film evaporation is achieved if the unbroken vapor film is widespread.

This occurrence is known as the Leidenfrost phenomenon.

In order to avoid this region in technical plants, the critical heat flux density must not be exceeded.

9.1.2 Flow boiling

In flow boiling, the evaporation occurs while streaming through tubes or the channels.

This application is common in technical plants, for example, in thermosiphon reboilers.

Due to the partial evaporation, a two-phase stream with lower density occurs and the flow velocity in the channels becomes strongly increased.

Consequently, the heat transfer coefficients get better.

Due to the heat input, bubbles evolve on the heating surface which get freed and reach the main streaming region. The streaming form of the two-phase stream changes its shape with increasing vapor composition from the bubble flow (5% vapor) to plug flow and to annular flow.

In technical reboilers, the plug flow and annular flow with high flow velocity and a liquid film on the heated wall should be strived for.

With increasing vaporization, the liquid film on the wall disappears and droplet flow occurs.

In this region, the heat transfer coefficient decreases dramatically because the heat transportation must take place through the poorly conducting vapor layer on the wall.

The region of the droplet flow can be avoided if the critical vapor rate is determined in which the transition from annular flow in droplet flow occurs.

No droplet flow occurs for vapor rates <30% in the evaporator outlet.

A special type of flow boiling is the falling film evaporation in which the evaporation occurs from a thin liquid film on the wall. This kind of evaporation is especially suitable for vacuum evaporations because the bubble point is not increased by the static height of a liquid or of a two-phase mixture.

9.1.3 Flash evaporation

In the flash evaporation, the liquid in the evaporator is not vaporized but it is heated under pressure to a temperature higher than the bubble point and the evaporation process takes place thereafter at the pressure relief in the pressure-reducing valve.

The evaporation of the superheated liquid takes place at the expansion valve outlet. The latent heat for evaporation is drawn from the superheated liquid.

This kind of vaporization is particularly suitable in the vaporization of dirty products with high flow velocities in order to reduce the fouling of the heating surface.

9.2 EVAPORATOR CONSTRUCTION TYPES

The industrially mostly used separation operation for fluid mixtures is the distillation or rectification. In these separation operations, the required heat is provided by an evaporator at the bottoms of the fractionation columns.

These evaporators in the bottoms of distillation columns are known as reboilers.

As opposed to the reboilers, solutions are concentrated in evaporators by vaporizing. The concentrate of the solution, for instance, sugar, can thereby be the required product or the evaporated exhaust vapor, for instance, water from sea water.

Reboiler types
In the following, the different reboiler construction types are treated:
- Vertical thermosiphon evaporators

- Horizontal thermosiphon evaporators
- Forced circulation evaporators
- Flash evaporators
- Kettle reboilers
- Internal evaporators
- Falling film evaporators

9.2.1 Thermosiphon evaporator

The process fluid flows through the evaporator from the bottom to the top.

As a result of the partial evaporation, a two-phase mixture with lower density occurs in the evaporator and due to the difference in static heights, natural circulation occurs. The driving pressure gradient must be greater than the pressure loss in the thermosiphon circle. It then comes to equilibrium between the driving height and the thermosiphon circulating rate.

A thermosiphon circle does not work,

- if the driving height for the thermosiphon circulation is too low
- if the driving temperature gradient is too low for a sufficient evaporation
- if the composition of the light boiling components in the mixture is too low
- if a large portion of the evaporator surface area is needed for the heating of the product up to the bubble point.
  Normal heat flux density: $q \approx 20{,}000{-}30{,}000 \text{ W/m}^2$
  Normal vaporization rate: 10–20% of the circulation at atmospheric pressure
  Normal vaporization rate: 30–50% of the circulation at vacuum evaporation
  Normal heat transfer coefficient at atmospheric pressure: $\alpha \approx 1500{-}2500 \text{ W/m}^2 \text{ K}$
  Normal heat transfer coefficient at vacuum evaporation: $\alpha \approx 1000{-}1500 \text{ W/m}^2 \text{ K}$
  The design of thermosiphon evaporators will be shown in Chapter 10.

9.2.1.1 Vertical thermosiphon circulation evaporator (Figure 9.1)

The evaporation occurs in the flow through vertical tubes that are heated on the shellside from the bottom to the top.

Tube lengths: 2 m (vacuum) to 4 m, normal 2.5–3 m
Tube diameter: 25–50 mm (vacuum)
Downcomer cross-section: 50% of the flow cross-section in the evaporator
Riser cross-section: 100% of the flow cross-section in the evaporator
Liquid height at atmospheric pressure: upper tube sheet
Liquid height at vacuum evaporation: 1 m under upper tube sheet.

Advantages

- Simple apparatus and easy to clean.
- Short residence time at the heating surface.

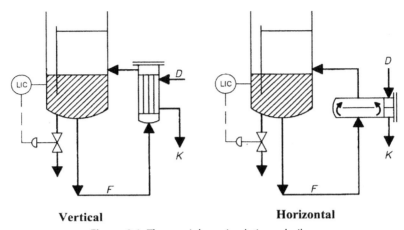

Figure 9.1 Thermosiphon circulation reboilers.

Disadvantages
- Bubble point increase because of the static height over the evaporator surface.
- The heat exchanger surface is constructively limited.
- Less suitable for vacuum evaporation.
- Not suitable for products with higher viscosity and dirty media and mixtures with wide boiling ranges which contain small amount of light boiling components.
- Pulsing stream at flow velocities in riser under 5 m/s.

9.2.1.2 Horizontal thermosiphon circulation evaporator (Figure 9.1)
The product is evaporated on the shellside and heated on the tubeside.

Recommendations
- Good distribution of the circulating liquid over several inlet nozzles.
- Sufficient flow velocity >1 m/s on the shellside.
- Control valve in the inlet to avoid over flooding.
- TEMA-construction types: AJL, AHL, AEL.

Advantages
- Lower bubble point increase than for vertical apparatus because the liquid height over the evaporator surface is less than in the vertical construction.
- No constructive dimension limitation of the heat exchanger surface.
- Less sensitive with regard to circulation for process fluctuations.
- Suitable for more viscous media up to 20 mPas.

Disadvantages
- Pull-through bundle with square pitch is necessary for cleaning.
- Difficult cleaning.

9.2.1.3 Once-through thermosiphon reboiler (Figure 9.2)

In once-through evaporator, the outflowing bottom liquid flows from the last bottom tray directly into the reboiler and the light components must be vaporized according to the specification in an evaporation cycle. This is more difficult for the production than with a circulating evaporator where the bottom rates are recirculated several times and hence the product specification of the bottom can be kept more easily through a larger buffer volume.

Advantages
- An additional theoretical stage for the separation.
- Low thermal damage because the product flows only once through the evaporator and stays very shortly in the hot zone.
- No bubble point increase by the high boiling components in the bottom.

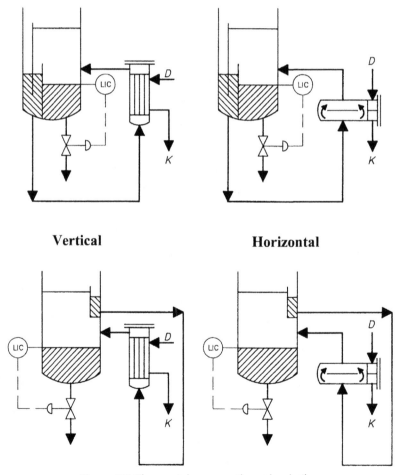

Vertical   **Horizontal**

Figure 9.2 Thermosiphon once-through reboiler.

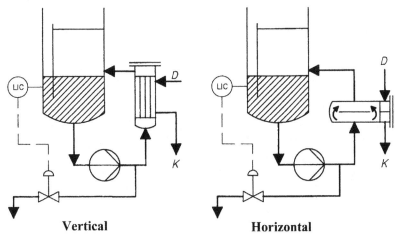

Vertical   **Horizontal**
Figure 9.3 Forced circulation reboiler.

Disadvantages
- Difficult product setting without circulation.

9.2.2 Forced circulation reboiler (Figure 9.3)

The product is pumped through the evaporator with a pump. This forced circulation is necessary if a thermosiphon circulation is not possible because the product is viscous or the product contains little volatile components in the vapor.

Heat flux density: $q \approx 30,000-50,000 \text{ W/m}^2$

Evaporation rate: 1–5% of the circulation

Determination of the α values according to the equations for the convective heat transfer for the corresponding flow velocity.

Advantages
- Suitable for high heat flow densities and dirty materials
- The circulation rate and the vaporization rate can be set individually
- Suitable for small evaporation rates.

Disadvantages
- High circulation rate necessary (energy costs)
- Circulating pump is required with sufficient net positive suction head (NPSH) value for boiling products
- Complex piping and danger of leakage in the pump.

9.2.3 Flash evaporator (Figure 9.4)

The liquid is heated under pressure in the evaporator.

The evaporation occurs after the flashing at a lower pressure in the reducing valve.

Large quantities circulated are necessary.

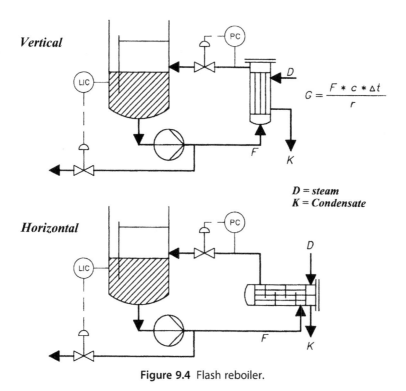

Figure 9.4 Flash reboiler.

Example 1: Calculation of the vapor rate G, which is released in flashing the heated circulated liquid F

F = circulating rate = 10, 000 kg/h c = spec. heat capacity = 2 kJ/kg K

r = latent heat = 250 kJ/kg Δt = heating temperature difference = 20 K

$$G = \frac{F \times c \times \Delta t}{r} = \frac{10,000 \times 2 \times 20}{250} = 1600 \text{ kg/h}$$

Recommended flow velocity in the tubes: 1.5–2.5 m/s.
The α value is determined with the equations for the convective heat transfer.

Advantages
Especially suitable for polymerizing or dirty products because the fouling danger can be reduced by high flow velocities in the tubes.

Disadvantages
Pump is necessary with sufficient NPSH value.

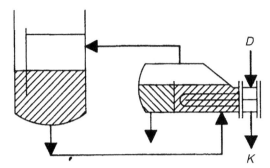

Figure 9.5 Kettle reboiler.

9.2.4 Kettle reboiler

This evaporator consists of a horizontal vessel with built-in tube bundle and an overflow weir in order that the tubes are constantly covered with the liquid (Figure 9.5).

Pure vapors flow back into the column, no two-phase mixture.

The liquid flows out over the weir and is drawn at the bottom. The inflow height to the evaporator must be sufficiently large.

Recommendations
Heat flux density: $q \approx 20,000-30,000$ W/m^2

Design: see "Dimensioning of kettle reboilers" in Section 9.3.2.

Vessel diameter: $\approx 1.5-2$ times heating bundle diameter.

Vapor velocity over tube bundle: <0.5 m/s.

Advantages
- No sensitive thermosiphon circle
- Simple design with one additional theoretical stage.

Disadvantages
- Heavy fouling and high boiling components accumulation
- Long residence time with the danger of thermal decomposition
- Difficult cleaning of the shellside
- Purge nozzle for dirt removing is necessary.

9.2.5 Internal evaporator (Figure 9.6)

In this construction, the tubes or plates heating bundle is built directly in the bottom of the column.

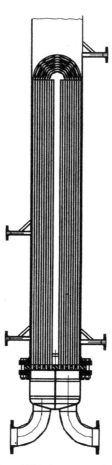

Figure 9.6 Internal U-tube evaporator in a column.

Advantages

*   An internal evaporator is cost-effective because there is no need for a vessel and piping
*   Lower fouling tendency than in kettle-type evaporator.

Disadvantages

*   The column diameter determines the heating bundle length and constrains the possible evaporator surface
*   For cleaning or repair, the column must be shut down, emptied, and gas freed.

9.2.6 Falling film evaporator

The liquid product is directed as irrigating liquid falling film on the inside of the tubes (Figure 9.7).

Cocurrent falling film reboiler

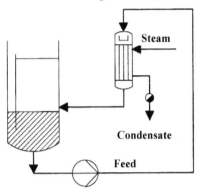

Countercurrent falling film reboiler

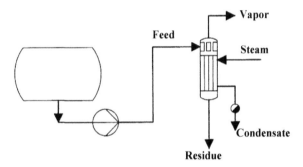

Figure 9.7 Falling film reboiler.

A good distribution is important. In order to avoid dry spaces, the irrigating rate of $1-2 \, m^3/h$ and m tube circumference must be kept. The liquid does not vaporize on the heating surface but on the liquid surface. No liquid height over the evaporator area exists and therefore no bubble point increase by a static height occurs.

This is why the falling film evaporator is especially suitable for evaporation in vacuum.

Recommended tube diameter: 40−80 mm, in order to minimize the pressure loss.

Advantages
- No bubble point increase and short residence times.

Disadvantages
- Complex construction with distributors and circulation pump is necessary
- Foulings by crack residues deteriorate the liquid distribution.

9.3 DESIGN OF EVAPORATORS FOR NUCLEATE BOILING [1−5]

In Figure 9.8, the *heat flux density q* of heated water as a function of the temperature difference, $T_W - T_B$, between the heated wall and the boiling liquid is shown. The following zones are noted:

At low temperature differences, no vaporization occurs but only a heating of the liquid by natural convection. This is the region up to point A. The heat transfer coefficients are small and lie in the range of 250−350 W/m² K.

The normal range of nucleate boiling lies between the points A and B.

The heat flux density increases strongly with increasing temperature difference, $T_W - T_B$.

The heat transfer coefficients for organic media lie in the range 2000−3000 W/m² K.

Some heat transfer coefficients are given in Figure 9.9.

The unproductive film boiling occurs for the large temperature differences in the range C to E.

The heat transfer coefficients are poor and due to the high wall temperatures, organic materials are decomposed.

The *heat transfer coefficients* for the vaporization are dependent on the temperature difference, $T_W - T_B$, and the physical properties of the medium being vaporized. In Figure 9.9, the heat transfer coefficients of some materials for the nucleate boiling as a function of the temperature difference between the heating wall and medium being vaporized are given.

The curves are valid for the boiling of single components at normal pressure.

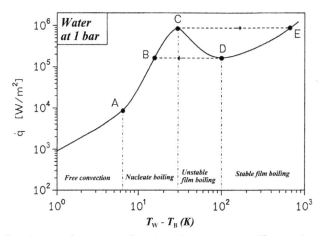

Figure 9.8 Heat flux density of water as a function of temperature difference between heating wall and the boiling medium.

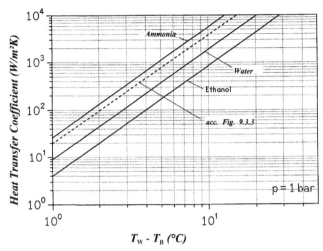

Figure 9.9 Heat transfer coefficient for the nucleate boiling as a function of the temperature difference between the heating wall and the boiling medium.

In evaporation in *vacuum*, the heat transfer coefficients are clearly less because the vapor bubbles are larger and cover more heating surface. The *heat transfer coefficients of mixtures* are smaller because the high boiling components increase the bubble point on the heating surface. The mixture must be heated in the evaporation from the bubble point up to the dew point.

In the equation of Mostinski [1], the different influences on the *heat transfer coefficient in nucleate boiling at a single tube* α_{NS} are covered:

$$\alpha_{NS} = 35.5 \times 10^{-6} \times P_c^{0.69}(Pa) \times q^{0.7} \times A \times B \left(W/m^2\,K\right)$$
$$\alpha_{NS} = 0.00417 \times P_c^{0.69}(kPa) \times q^{0.7} \times A \times B \left(W/m^2\,K\right)$$

P_c = critical pressure (Pa or kPa)
A = correction factor for the boiling pressure
B = correction factor for mixtures ($B = 1$ for single components)
Calculation of the correction factor A for the pressure and B for mixtures with a boiling range, BR:

$$A = 1.8 \times P_R^{0.17} + 4 \times P_R^{1.2} \qquad B = \frac{1}{1 + 0.027 \times q^{0.1} \times BR^{0.7}}$$

BR = boiling range (K) = temperature difference between dew and boiling point
P_c = critical pressure (bar)
P_R = reduced pressure = P/P_c
P = working pressure (bar)
q = heat flux density (W/m^2)

Example 2: Calculation of the heat transfer coefficient of boiling benzene on a single tube at different pressures

$$P_C = 48.9 \text{ bar} \quad q = 30{,}000 \text{ W/m}^2 \quad B = 1 \text{ (single component)}$$

$$\alpha_{NS} = 0.00417 \times 4890^{0.69} \times 30{,}000^{0.7} \times A = 1994 \times A$$

$$P = 1 \text{ bar}$$

$$P_R = \frac{1}{48.9} = 0.0204 \qquad A = 1.8 \times 0.0204^{0.17} + 4 \times 0.0204^{1.2} = 0.967$$

$$\alpha_{NS} = 0.987 \times 1994 = 1927 \text{ W/m}^2 \text{ K}$$

$$P = 0.1 \text{ bar}$$

$$P_R = \frac{0.1}{48.9} = 0.00204 \qquad A = 1.8 \times 0.00204^{0.17} + 4 \times 0.00204^{1.2} = 0.63$$

$$\alpha_{NS} = 0.63 \times 1994 = 1257 \text{ W/m}^2 \text{ K}$$

In vacuum at 100 mbar, the heat transfer coefficient is clearly less than at 1 bar.

Example 3: Calculation of the heat transfer coefficient of boiling ethanol on a single tube at 1 bar and at different heat flux densities

$$P_C = 63.8 \text{ bar} = 6380 \text{ kPa} \quad P_R = 1/63.8 = 0.0157 \quad A = 0.915 \quad B = 1$$

$$q = 10{,}000 \text{ W/m}^2$$

$$\alpha_{NS} = 0.00417 \times 6380^{0.69} \times 10{,}000^{0.7} \times 0.915 = 1016 \text{ W/m}^2 \text{ K}$$

$$q = 20{,}000 \text{ W/m}^2$$

$$\alpha_{NS} = 0.00417 \times 6380^{0.69} \times 20{,}000^{0.7} \times 0.915 = 1650 \text{ W/m}^2 \text{ K}$$

$$q = 30{,}000 \text{ W/m}^2$$

$$\alpha_{NS} = 1.6102 \times 30{,}000^{0.7} = 2192 \text{ W/m}^2 \text{ K}$$

$$q = 60{,}000 \text{ W/m}^2$$

$$\alpha_{NS} = 1.6102 \times 60{,}000^{0.7} = 3561 \text{ W/m}^2 \text{ K}$$

The heat transfer coefficient becomes greater with increasing heat flux density.

Example 4: Calculation of the heat transfer coefficient for a benzene–toluene mixture with a boiling range, BR = 10 °C

$$\alpha_{NS} = 1927 \text{ W/m}^2 \text{ K from Example 1 for benzene.}$$

$$B = \frac{1}{1 + 0.027 \times 30{,}000 \times 10^{0.7}} = 0.725$$

$$\alpha_{NS} = 0.725 \times 1927 = 1397 \text{ W/m}^2 \text{ K}$$

The heat transfer coefficient for the mixture is smaller *than the value for the single component*.

Heat transfer coefficient α_{NB} in the tube bundle

The heat transfer coefficient α_{NB} in the tube bundle is clearly better than the heat transfer coefficient on a single tube α_{BS} due to the larger turbulence. With the correction factor $F_B = 1.5$ for the tube bundle, the following equation results:

$$\alpha_{NB} \approx 1.5 \times \alpha_{NS} = 0.00626 \times P_c^{0.69} \times q^{0.7} \times A \times B \; \left(W/m^2 \; K\right)$$

Example 5: Calculation of the heat transfer coefficient of the boiling benzene–toluene mixture from Example 4 in a tube bundle

$$\alpha_{NB} = 1.5 \times 1397 = 2095 \; W/m^2 \; K$$

Maximum allowable heat flux density in the tube bundle [6,7]

The maximum allowable heat flux density for a tube bundle evaporator results from the allowable vapor velocity over the projected bundle surface area A_P.

Too high bubble velocities give rise to a vapor bubble congestion in the bundle. A vapor bubble film then disturbs the heat transfer. Film boiling occurs with poor heat transfer coefficients.

For the allowable *vapor rising velocity*, w_{allow}, over a tube bundle or a kettle-type evaporator in order to avoid bubble congestion between the tubes, the following holds:

$$w_{allow} = 0.18 \times \left(\frac{\left(\rho_{liq} - \rho_V\right) \times \sigma \times 9.81}{\rho_V^2}\right)^{0.25} \; (m/s)$$

w_{allow} = flow velocity based on the projection of the bundle (m/s)
ρ_V = vapor density (kg/m^3)
ρ_{liq} = liquid density (kg/m^3)
σ = surface tension (N/m)

From the allowable vapor stream velocity, w_{allow}, the practically allowable heat flux density without vapor bubble congestion between the tubes can be calculated.

$$q_{allow} = \frac{w_{allow} \times A_P \times 3600 \times \rho_V \times r}{A} \; \left(W/m^2\right)$$

A = heat exchanger-heat transfer surface area (m^2)
$A_P = d_B \times L_B$ = projected bundle surface area (m^2)
d_B = bundle width (m)
L_B = bundle length (m)
r = heat of vaporization (Wh/kg)

Example 6: Determination of the allowable vapor velocity over the bundle and the allowable heat flux density for avoiding bubble congestion

Bundle diameter $d_B = 0.325$ m   Bundle length $L_B = 4$ m
$A_P = 0.325 \times 4 = 1.3$ m^2 $A = 21.2$ m^2
$\sigma = 0.0155$ N/m $r = 234$ Wh/kg $\rho_V = 1.57$ kg/m^3 $\rho_{liq} = 719$ kg/m^3

$$w_{allow} = 0.18 \times \left(\frac{(719 - 1.57) \times 0.0155 \times 9.81}{1.57^2} \right)^{0.25} = 0.46 \text{ m/s}$$

$$q_{allow} = \frac{0.46 \times 1.3 \times 3600 \times 1.57 \times 234}{21.2} = 37,652 \text{ W/m}^2$$

9.3.1 Practical design with examples

The heat transfer coefficients can be very easily determined with the diagram according to Kern [5] given in Figure 9.10. The heat transfer coefficients for the *nucleate boiling* and the heating by natural *convection* are shown as function of the temperature difference, $T_W - T_B$.

In practice, the *heat flux density q is limited* in order to minimize the product damage in the evaporator and to avoid vapor bubble congestion.

$$q = \frac{Q}{A} \; (\text{W/m}^2)$$

$Q =$ heat load (W)
$A =$ heating area (m^2)
Recommendation for water and aqueous solutions:
$q_{max} \approx 90{,}000$ bis $120{,}000$ W/m^2
$\alpha_{NB} \approx 5500$ bis $11{,}000$ W/m^2 K
Recommendation for organic media:
Forced circulation: $q_{max} \approx 60{,}000 - 95{,}000$ W/m^2
Natural circulation: $q_{max} \approx 30{,}000 - 50{,}000$ W/m^2
$\alpha_{NB} \approx 1700 - 2500$ W/m^{2}K

Often the product must be first heated to the boiling temperature before the real evaporation. This is the case especially for product mixtures with a boiling range. In such cases, an average weighted heat transfer coefficient, α_a, is used for the supply of sensible heat by natural convection and the evaporation. This average heat transfer coefficient is determined as follows:

$$\alpha_a = \frac{Q_{tot}}{\dfrac{Q_H}{\alpha_{heating}} + \dfrac{Q_E}{\alpha_{boiling}}} \; (\text{W/m}^2 \text{ K})$$

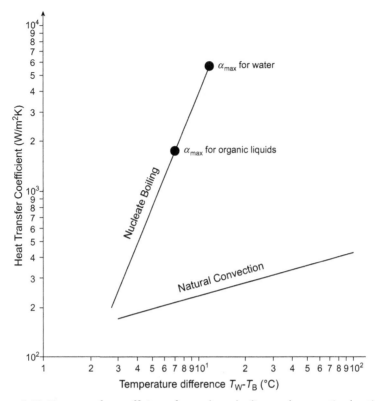

Figure 9.10 Heat transfer coefficients for nucleate boiling and convective heating.

Q_H = heat load for convective heating (W)
Q_E = heat load for the evaporation (W)
$Q_{tot} = Q_H + Q_E$ (W)
$\alpha_{boiling}$ = heat transfer coefficient for the evaporation (W/m^2 K)
$\alpha_{heating}$ = heat transfer coefficient for the heating by natural convection (W/m^2 K)

Example 7: Overall heat transfer coefficient for heating up and evaporation of gasoline

Feed to the evaporator: 17,500 kg/h   Evaporated rate: 13,000 kg/h
Heating in the evaporator from 157 to 168 °C
Evaporation: $\alpha_{boiling}$ = 2500 W/m^2 K
Heating: $\alpha_{heating}$ = 335 W/m^2 K
Evaporation load: Q_E = 1060 kW   Heating load: Q_H = 150 kW

$$\alpha_a = \frac{150 + 1060}{\dfrac{150}{335} + \dfrac{1060}{2500}} = 1388 \text{ W/m}^2 \text{ K}$$

This example makes clear how strong the heat transfer coefficient for the nucleate boiling is reduced by the heating process.

Guide line figures for the overall heat transfer coefficients of steam-heated reboilers:

Light hydrocarbons: $U \approx 900-1200$ W/m^2 K

Heavy hydrocarbons: $U \approx 550-900$ W/m^2 K

Water and aqueous solutions: $U \approx 1250-2000$ W/m^2 K

Guide line figures for thermal oil-heated reboilers: $U \approx 400-600$ W/m^2 K

The *overall heat transfer coefficient U* of evaporators is predominantly determined by the fouling of the heating surfaces.

Example 8: Steam-heated kettle-type evaporator

$\alpha_{io} = 5000$ W/m^2 K for the steam condensation in the tubes.

Fouling factor for both sides $f_{tot} = f_i + f_a = 0.0004$.

$\alpha_{NB} = 2000$ W/m^2 K for the evaporation outside the tubes with $f_{tot} = 0.0004$

$$\frac{1}{U} = \frac{1}{5000} + \frac{1}{2000} + 0.0004 \qquad U = 909 \text{ W/m}^2 \text{ K}$$

$\alpha_{NB} = 2000$ W/m^2 K for the evaporation outside the tubes with $f_{tot} = 0.001$

$$\frac{1}{U} = \frac{1}{5000} + \frac{1}{2000} + 0.001 \qquad U = 588 \text{ W/m}^2 \text{ K}$$

Example 9: Design of a tube bundle for a distillation still

Determination of the necessary heating area for a tube bundle of 25 × 2 tubes.

Required heat load $Q = 850,000$ W for the distillation.

Bubble point temperature $T_S = 107\,°C$ Heating steam temperature $T_{steam} = 140\,°C$

$\Delta t = 140-107 = 33\,°C$

Calculation of the wall temperature T_W and of the overall heat transfer coefficient U:

$\alpha_i = 6000$ W/m^{2}K for the heating with condensing steam in the 25 × 2 tubes

$$\alpha_{io} = \alpha_i \times \frac{d_i}{d_o} = 6000 \times \frac{21}{25} = 5040 \text{ W/m}^2 \text{ K} \quad \text{based on the tube outside}$$

$$\alpha_{iof} = \frac{1}{\dfrac{1}{5040} + 0.0001} = 3351 \text{ W/m}^2 \text{ K} \quad \text{with fouling}$$

Estimated value for the heat transfer by nucleate boiling: $\alpha_{NB} = 1700$ W/m^2 K on the tube outside.

For the determination of α_{NB} from Figure 9.10, the difference between the wall and the boiling temperature, T_W-T_B, is needed:

$$T_W = T_B + \frac{\alpha_{iof}}{\alpha_{iof} + \alpha_{NB}} \times (T_{steam} - T_B) = 107 + \frac{3351}{3351 + 1700} \times (140 - 107) = 128.9\,°C$$

$$T_W - T_B = 128.9 - 107 = 21.9\,°C$$

From Figure 9.10 for $T_W - T_B = 21.9\,°C$, a clearly higher value results than the recommended maximum value of $\alpha_{NB} \approx 2500\ \text{W/m}^2\ \text{K}$.

Chosen: $\alpha_{NB} = 2000\ \text{W/m}^2\ \text{K}$

The overall heat transfer coefficient U is derived as follows:

$$\frac{1}{U} = \frac{1}{\alpha_{iof}} + \frac{1}{\alpha_{NB}} + \frac{s}{\lambda} + f_{boiling}$$

α_{iof} = heat transfer coefficient in the tube with fouling = 3351 (W/m² K)
α_{NB} = heat transfer coefficient for the boiling = 2000 (W/m² K)
s = tube wall thickness = 0.002 (m)
λ = thermal heat conductivity of the tube material = 14 (W/m K)
$f_{boiling}$ = fouling factor on the vaporization side = 0.0002 (m² K/W)

$$\frac{1}{U} = \frac{1}{3351} + \frac{1}{2000} + \frac{0.002}{14} + 0.0002 = \frac{1}{876} \qquad U = 876\ \text{W/m}^2\text{K}$$

Hence the following required heating area A_{req} results:

$$A_{req} = \frac{Q}{U \times \Delta t} = \frac{850,000}{876 \times 33} = 29.4\ \text{m}^2$$

The heat flux density is: $q = U \times \Delta t = 876 \times 33 = 28,908\ \text{W/m}^2$

The recommended heat flux densities for the different evaporator types should preferably be maintained in order not to thermally damage the product.

The temperature difference, $T_W - T_B$, between the heating wall and the boiling liquid should be in the order of ca. 20 °C.

Example 10: Butane evaporation at 20 bar in an evaporator with 33 m² area

151 tubes 15.6/19 mm diameter, 3.66 m long.
Evaporated butane rate W = 18,523 kg/h

$T_B = 117\,°C$	$T_{steam} = 157\,°C$	$P = 20$ bar	$P_C = 37.6$ bar
$\rho_{liq} = 433\ \text{kg/m}^3$	$\rho_V = 36.4\ \text{kg/m}^3$	$c = 0.73$ Wh/kg K	
$\sigma = 0.003$ N/m	$\lambda = 0.09$ W/m K	$\eta = 0.1$ mPas	$r = 62$ Wh/kg

Heat load Q and heat flux density q:

$$Q = 18,523 \times 62 = 1,148,426\ \text{W}$$

$$q = \frac{114,426}{33} = 34,800\ \text{W/m}^2$$

Calculation of the heat transfer coefficient, α_{NB}, according to Mostinski for a heating bundle with pressure correction factor A:

$$\alpha_{NB} = 0.00626 \times 3760^{0.69} \times 34,800^{0.7} \times A = 2767.5 \times A$$

$$A = 1.8 \times \left(\frac{20}{37.6}\right)^{0.17} + 4 \times \left(\frac{20}{37.6}\right)^{1.2} = 3.49$$

$$\alpha_{NB} = 3.49 \times 2767.5 = 9658.6\ \text{W/m}^2\ \text{K}$$

Check of the maximum allowable heat flux density q_{allow}:

$$w_{allow} = 0.18 \times \left(\frac{(\rho_{liq} - \rho_V) \times \sigma \times 9.81}{\rho_V^2}\right)^{0.25} = 0.18 \times \left(\frac{(433 - 36.4) \times 0.003 \times 9.81}{36.4^2}\right)^{0.25}$$

$$= 0.055 \text{ m/s}$$

Projected bundle surface area A_P: $A_P = d_B \times L_B = 0.4 \times 3.66 = 1.46 \text{ m}^2$
Maximum allowable heat flux density q_{allow}:

$$q_{allow} = \frac{w_{allow} \times A_P \times 3600 \times \rho_V \times r}{A} = \frac{0.055 \times 1.46 \times 3600 \times 36.4 \times 62}{33} = 19,876 \text{ W/m}^2$$

Allowable butane evaporation W_{allow}:

$$W_{allow} = q_{allow} \times \frac{A}{r} = 19,876 \times \frac{33}{62} = 10,579 \text{ kg/h}$$

18,523 kg/h cannot be evaporated but only 10,579 kg/h.

9.3.2 Dimensioning of kettle reboilers [5−7]

9.3.2.1 Hydraulic design

For the operation of a kettle-type reboiler, the required height ΔH between the liquid height in the column and the overflow weir in the reboiler must be given (Figure 9.11). If the height ΔH is not sufficiently dimensioned, the liquid height in the bottom of the column rises up to the maximum of the height of the vapor returning nozzle in which case the function of the kettle-type reboiler is blocked.

The necessary height ΔH results from the pressure losses.

Calculation of the required height ΔH with the security factor 2:

$$\Delta H = \frac{2 \times \left(\Delta P_{liq} + \Delta P_V + \Delta P_A\right)}{g \times \rho_{liq}} \text{ (m static head)}$$

$$\Delta P_{liq} = \left(f \times \frac{L}{d} + K_{liq}\right) \times \frac{w_{liq}^2 \times \rho_{liq}}{2} \text{ (Pa)}$$

$$\Delta P_V = \left(f \times \frac{L}{d} + K_V\right) \times \frac{w_V^2 \times \rho_V}{2} \text{ (Pa)}$$

$$\Delta P_A = \frac{m_V^2}{\rho_V} - \frac{m_{liq}^2}{\rho_{liq}} \approx \frac{m_V^2}{\rho_V} \text{ (Pa)}$$

$$m_V = w_V \times \rho_V \text{ } (\text{kg/m}^2 \text{ s}) \qquad m_{liq} = w_{Reb} \times \rho_{liq} \text{ } (\text{kg/m}^2 \text{ s})$$

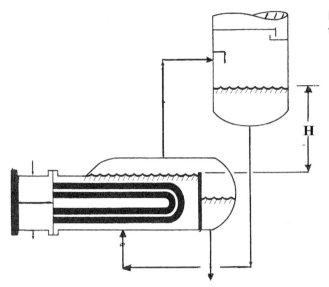

Figure 9.11 Required height difference ΔH for a kettle reboiler

H

ΔP_{liq} = pressure loss in the liquid line from the column to the reboiler (Pa)
ΔP_V = pressure loss in the vapor line from the reboiler to the column (Pa)
ΔP_A = acceleration pressure loss in the reboiler (Pa)
G_V = vapor rate to the column (kg/h)
G_{liq} = total liquid rate to the reboiler (kg/h)
G_S = liquid draw-off rate from the reboiler (kg/h)
m_V = mass flux density of the vapors in the vapor line (kg/m² s)
m_{liq} = mass flux density of the liquid in the reboiler (kg/m² s)
f = friction factor of the piping
L = piping length (m)
K_{liq} = resistance factors of the fittings in the liquid line
K_V = resistance factors of the fittings in the vapor line
w_V = flow velocity in the vapor line (m/s)
w_{liq} = flow velocity in piping to the reboiler (m/s)
w_{Reb} = liquid velocity in the reboiler (m/s)
ρ_V = vapor density (kg/m³)
ρ_{Fl} = liquid density (kg/m³)

Example 11: Calculation of the required height difference for a kettle reboiler

$G_{\text{liq}} = 20$ t/h	$G_V = 15$ t/h	$G_S = 5$ t/h	$K_{\text{liq}} = 3$	$K_V = 2$
$\rho_{\text{liq}} = 750$ kg/m³	$V_{\text{liq}} = 26.7$ m³/h	$d = 0.1$ m	$w_{\text{liq}} = 0.94$ m/s	$L = 10$ m
$\rho_V = 5$ kg/m³	$V_V = 3000$ m³/h	$d = 0.25$ m	$w_V = 17$ m/s	$L = 10$ m
$w_{\text{Reb}} = 0.01$ m/s				

$$\Delta P_{\text{liq}} = \left(0.04 \times \frac{10}{0.1} + 3\right) \times \frac{0.94^2 \times 750}{2} = 2319 \text{ Pa}$$

$$\Delta P_V = \left(0.04 \times \frac{10}{0.25} + 2\right) \times \frac{17^2 \times 5}{2} = 2601 \text{ Pa}$$

$$m_V = 17 \times 5 = 85 \text{ kg/m}^2 \text{ s} \quad m_{\text{liq}} = 0.01 \times 750 = 7.5 \text{ kg/m}^2 \text{ s}$$

$$\Delta P_A = \frac{85^2}{5} = 1445 \text{ Pa}$$

$$\Delta H = \frac{2 \times (2319 + 2601 + 1445)}{9.81 \times 750} = 1.73 \text{ m}$$

What is to be considered?

1. The vapor height above the liquid should be at least 0.5 m in order that preferably no entrainment occurs. The entrained droplets increase the vapor density and hence the pressure drop in the vapor line ΔP_V.
2. The weir overflow height must be considered in establishing the vapor height because the free height above the liquid is thereby reduced.
3. A good distribution of the liquid rate entering the reboiler at the bottom without resistances at the inlet in the reboiler, for instance, in restrictions by the support for the tube bundle which increase the liquid pressure loss ΔP_{liq}.
4. The liquid should be introduced over two or three inlet nozzles in order to preferably achieve uniform evaporation.

9.3.2.2 Thermal design

First the heat transfer coefficient, α_{NB}, is determined for the nucleate boiling.

Calculation of the *heat transfer coefficient, α_{NB}, according to Mostinski* for nucleate boiling or pool boiling in the tube bundle with the correction factors A for the pressure and B for the boiling range in product mixtures:

$$\alpha_{\text{NB}} = 0.00626 \times P_c^{0.69} \times q^{0.7} \times A \times B \ \left(\text{W}/\text{m}^2 \text{ K}\right)$$

Calculation of the pressure correction factor A:

$$A = \text{pressure factor} = 1.8 \times P_R^{0.17} + 4 \times P_R^{1.2}$$

Calculation of the correction factor B for mixtures with the boiling range, BR ($^\circ$C):

$B = \exp\left(-0.027 \times \text{BR}\right)$

P_c = critical pressure (kPa)

q = heat flux density (W/m^2)

Example 12: Calculation of a steam-heated kettle-type evaporator

$$q = 33,000 \text{ W/m}^2 \qquad P_c = 3900 \text{ kPa} \qquad P = 1 \text{ bar} \qquad BR = 10\,°C$$
$$B = \exp(-0.027 \times 10) = 0.76$$

$$P_R = \frac{100}{3900} = 0.0256 \quad A = 1.8 \times 0.0256^{0.17} + 4 \times 0.0256^{1.2} = 1.015$$

$$\alpha_{NB} = 0.00626 \times 3900^{0.69} \times 33,000^{0.7} \times 1.015 \times 0.76 = 2111 \text{ W/m}^2\,K$$

Calculation of the overall heat transfer coefficient:

$\alpha_{io} = 8000 \text{ W/m}^2K$ for condensing steam.

$$s = 2 \text{ mm} \qquad \lambda_{Wall} = 50 \text{ W/m K} \qquad \text{Fouling factor } f = 0.0002$$
$$\frac{1}{U} = \frac{1}{2111} + \frac{1}{8000} + \frac{0.002}{50} + 0.0002 \quad U = 1192 \text{ W/m}^2\,K$$
$$Q = 0.7 \text{ MW} \quad LMTD = 22\,°C$$
$$A = \frac{Q}{U \times LMTD} = \frac{700,000}{1192 \times 22} = 26.7 \text{ m}^2$$

9.3.2.3 *Allowable bubble rising velocity [6,7]*

If the heat flux density is too high, vapor bubble congestion between the tubes of the kettle-type evaporator occurs. This results in an undesired film evaporation because the vapor bubbles block the heating area. The danger of vapor bubble congestion can be reduced by using a square tube arrangement with large pitch.

In the design of a kettle-type evaporator, the maximum allowable rising velocity, w_{allow}, of the evaporated vapors over the projected area of the evaporator bundle A_P should not be exceeded.

$$A_P = d_B \times L_B = \text{bundle width} \times \text{bundle length}$$

The maximum vapor rising velocity, w_{allow}, above the projected area A_P is determined as follows:

$$w_{allow} = 0.18 \times \left(\frac{\left(\rho_{liq} - \rho_V^*\right) \times \sigma \times 9.81}{\rho_V^2} \right)^{0.25} \text{(m/s)}$$

$\rho_{liq} = $ liquid density (kg/m^3)
$\rho_V = $ vapor density (kg/m^3)
$\sigma = $ surface tension (N/m)

9.3.2.4 Allowable heat flux density in the kettle reboiler

The allowable bubble rising velocity limits the allowable heat flux density of the evaporator surface area A as follows:

$$q_{allow} = \frac{w_{allow} \times A_P \times 3600 \times \rho_V \times r}{A} \ (W/m^2)$$

$r =$ heat of vaporization (Wh/kg)

Example 13: Calculation of the allowable bubble rising velocity and the heat flux density for a kettle-type evaporator

$\rho_{liq} =$ liquid density $= 775$ kg/m^3 $\rho_V =$ vapor density $= 3$ kg/m^3
$\sigma =$ surface tension $= 0.018$ N/m $r = 86.4$ Wh/kg $A = 31.6$ m^2
$d_B = 0.45$ m $L = 6$ m

$$w_{allow} = 0.18 \times \left(\frac{(775 - 3) \times 0.018 \times 9.81}{3^2}\right)^{0.25} = 0.355 \ m/s$$

$$Pojected \ area \ A_P = d_B \times L = 0.45 \times 6 = 2.7 \ m^2$$

$$Vapor \ rate \ V = 2700 \ m^3/h$$

$$Bubble \ velocity \ w_{bubble} = \frac{2700}{3600 \times 2.7} = 0.277 \ m/s$$

The real rising velocity of the vapor bubbles lies with 0.277 m/s under the maximum allowable bubble rising velocity of 0.355 m/s.

Allowable heat flux density due to the bubble rising velocity:

$$q_{allow} = \frac{0.355 \times 2.7 \times 3600 \times 3 \times 86.4}{31.6} = 28,303 \ W/m^2$$

Calculation of the real heat flux density q for the heat load $Q = 0.7$ MW:

$$q = \frac{700,000}{31.6} = 22,152 \ W/m^2$$

The available heat flux density of $q = 22,152$ W/m^2 is less than the allowable heat flux density $q_{zul} = 28,303$ W/m^2.

Example 14: Ethanol evaporation with the kettle-type evaporator

Data: $A = 31.6$ m^2 $F_B = 2.7$ m^2 $\rho_{liq} = 757$ kg/m^3 $\rho_V = 1.435$ kg/m^3
         $r = 267$ Wh/kg $\sigma = 17.7$ mN/m $P_c = 6390$ kPa $q = 30,000$ W/m^2

Calculation of the heat transfer coefficient, α_{NB}, for nucleate boiling:

$$\alpha_{NB} = 0.00626 \times 6390^{0.69} \times 30,000^{0.7} = 3600 \ W/m^2 \ K$$

Calculation of the overall heat transfer coefficient:

$\alpha_i = 8000 \text{ W/m}^2 \text{ K} \quad s = 2 \text{ mm} \quad \lambda_{Wall} = 50 \text{ W/m K} \quad f_{tot} = 0.0003 \quad A = 31.6 \text{ m}^2$
$\text{LMTD} = 22 \,°\text{C}$

$$\frac{1}{U} = \frac{1}{3600} + \frac{1}{8000} + \frac{0.002}{50} + 0.0003 \qquad U = 1346 \text{ W/m}^2 \text{ K}$$

Heat flux density $q = U \times \text{LMTD} = 1346 \times 22 = 29{,}620 \text{ W/m}^2$

Heat duty $Q = q \times A = 29{,}620 \times 31.6 = 935{,}988 \text{ W}$

$$\text{Vapor rate } V = \frac{Q}{r \times \rho_V} = \frac{935{,}988}{267 \times 1.435} = 2442.9 \text{ m}^3/\text{h}$$

$$\text{Bubble velocity } w_{bubble} = \frac{2442.9}{2.7 \times 3600} = 0.25 \text{ m/s}$$

Calculation of the allowable vapor rising velocity, w_{allow}

$$w_{allow} = 0.18 \times \left(\frac{(757 - 1.435) \times 0.0177 \times 9.81}{1.435^2}\right)^{0.25} = 0.51 \text{ m/s} \quad > w_{bubble} = 0.25 \text{ m/s}$$

Calculation of the allowable heat flux density, q_{allow}

$$q_{allow} = \frac{0.51 \times 2.7 \times 3600 \times 1.435 \times 267}{31.6} = 59{,}933 \text{ W/m}^2 \quad > q = 29{,}620 \text{ W/m}^2$$

Example 15: Water evaporation with the kettle-type evaporator

Water data:
$\rho_{liq} = 958 \text{ kg/m}^3 \qquad \rho_V = 0.597 \text{ kg/m}^3 \qquad r = 626.9 \text{ Wh/kg}$
$\sigma = 58.9 \text{ mN/m} \qquad P_c = 22{,}129 \text{ kPa} \qquad Q = 40{,}000 \text{ W/m}^2$
$A = 31.6 \text{ m}^2 \qquad \text{LMTD} = 22\,°\text{C} \qquad f_{tot} = 0.0003$
$s = 2 \text{ mm} \qquad \lambda = 50 \text{ W/m K} \qquad \alpha_{steam} = 8000 \text{ W/m}^2 \text{ K}$
$A_P = 2.7 \text{ m}^2$

$$\alpha_{NB} = 0.00626 \times 22{,}129^{0.69} \times 40{,}000^{0.7} = 10{,}376 \text{ W/m}^2 \text{ K}$$

$$\frac{1}{U} = \frac{1}{10{,}376} + \frac{1}{8000} + \frac{0.002}{50} + 0.0003 \qquad U = 1781 \text{ W/m}^2 \text{ K}$$

$q = k \times \text{LMTD} = 1781 \times 22 = 39{,}289 \text{ W/m}^2 \qquad Q = q \times A = 39{,}189 \times 31.6 = 1.238 \text{ MW}$

$$V = \frac{1.238 \times 10^6}{626.9 \times 0.597} = 3308.9 \text{ m}^3/\text{h} \qquad w_{bubble} = \frac{3308.9}{2.7 \times 3600} = 0.34 \text{ m/s}$$

Calculation of the allowable bubble rising velocity and the allowable heat flux density:

$$w_{allow} = 0.18 \times \left(\frac{(958 - 0.597) \times 0.0589 \times 9.81}{0.597^2}\right)^{0.25} = 1.13 \text{ m/s} \quad > w_{bubble} = 0.34 \text{ m/s}$$

$$q_{allow} = \frac{1.13 \times 2.7 \times 3600 \times 0.597 \times 626.9}{31.6} = 130{,}086 \text{ W/m}^2 \quad > q = 39{,}189 \text{ W/m}^2$$

9.3.2.5 Entrainment

In order to avoid too much entrainment of droplets in the rising vapor, the shell diameter D_M of the kettle-type evaporator is enlarged by the factor 1.5 to 2 and is greater than the heating bundle diameter d_B.

$$D_M \approx 1.5 \times d_B$$

The required flow cross-section A_{req} above the boiling liquid to avoid the droplet entrainment is determined from the allowable vapor velocity, $w_{droplet}$:

$$w_{droplet} = 0.04 \times \sqrt{\frac{\rho_{liq}}{\rho_V} - 1} \ (m/s)$$

$$A_{req} = \frac{V_{Vapor}}{w_{droplet}} \ (m^2) \qquad L_{req} = \frac{A_{req}}{D_M} \ (m)$$

V = vapor rate (m^3/s) L_{req} = required shell vessel length (m)
ρ_{liq} = liquid density (kg/m^3) ρ_V = vapor density (kg/m^3)

Example 16: Calculation of the required flow cross-section against droplet entrainment

$$V = 3000 \ m^3/h \qquad \rho_{liq} = 580 \ kg/m^3 \qquad \rho_V = 6 \ kg/m^3 \qquad d_B = 0.4 \ m \qquad D_M = 0.8 \ m$$

$$w_{droplet} = 0.04 \times \sqrt{\frac{580}{6} - 1} = 0.39 \ m/s$$

$$A_{req} = \frac{3000}{3600 \times 0.39} = 2.13 \ m^2$$

$$L_{req} = \frac{2.13}{0.8} = 2.66 \ m$$

9.4 DESIGN OF FALLING FILM EVAPORATORS [8–10]

In evaporation in vacuum, falling film evaporators are used because in these equipments not the bubble point is raised by a static liquid height. For a given heat duty Q, initially, the required evaporator area A is determined with an estimated overall heat transfer co-efficient U for the effective temperature difference Δt.

$$A = \frac{Q}{U \times \Delta t} \ (m^2)$$

Then a number n of the required tubes for a chosen tube dimension is established and the maximum vapor velocity w_V in the tube outlet is calculated.

$$n = \frac{A}{d_o \times \pi \times L}$$

$$w_V = \frac{Q}{r \times a_t \times \rho_V \times 3600} \ (m/s)$$

$$a_t = n \times d_i^2 \times \frac{\pi}{4} \ (m^2)$$

d_o = tube outer diameter (m) d_i = tube inner diameter (m)
L = tube length (m) a_t = tubes flow cross-section (m^2)
ρ_V = vapor density (kg/m^3) A = reboiler area (m^2)

The pressure drop in the tubes should be preferably low in vacuum vaporization in order to avoid a boiling point increase due to the higher pressure in the evaporator.

From Figure 9.12, it follows that the pressure loss for the toluene evaporation in vacuum increases with increasing vapor rate and that a higher liquid feed rate results in higher pressure loss.

Figure 9.13 shows how the pressure loss increases with the vapor rate in a hexadecane evaporator.

From Figure 9.13, it can be derived:
* that the pressure loss greatly increases with increasing vapor rate,
* that at a lower pressure, the pressure losses are greater,
* that a higher liquid feed increases the pressure loss.

In calculating the pressure losses for the two-phase stream in the evaporator, the increasing vapor fraction of the two-phase mixture due to the increasing tube length must be taken into account.

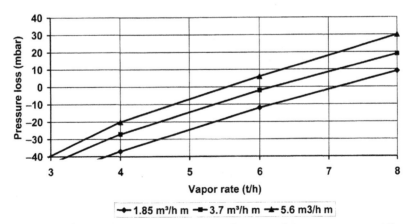

Figure 9.12 Pressure loss in a toluene evaporator as function of the vapor rate at different liquid feed rates.

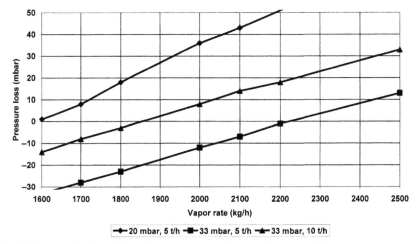

Figure 9.13 Pressure loss in a falling film evaporator for hexadecane at different pressures and liquid feeds as function of the vapor rate.

The total pressure drop in the evaporator is composed of three parts:

The friction pressure drop ΔP_{Fr}, the one that is increasing with the tube length.

The acceleration pressure drop ΔP_A, which increases with the evaporator length.

The negative hydrostatic pressure ΔP_H, which is increasing toward the end of the tube.

$$\Delta p_{tot} = \Delta P_{Fr} + \Delta P_A - \Delta P_H$$

In Figure 9.14, the function of the pressure drop composites in a falling film evaporator for hexadecane at a pressure of 33 mbar is shown.

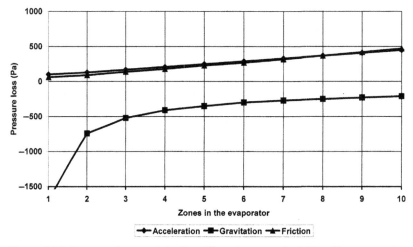

Figure 9.14 Pressure drop parts in the different zones of a falling film evaporator.

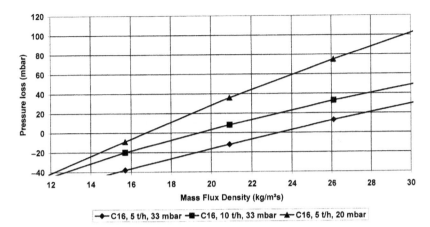

Figure 9.15 Pressure loss in a hexadecane evaporator as function of the mass flux density.

In order to avoid higher pressure losses, the mass stream density m in the evaporator tubes should be in the range of $15-18 \, \text{kg/m}^2 \, \text{s}$ maximum (see Figure 9.15).

$$m = w_V \times \rho_V \, (\text{kg/m}^2 \, \text{s})$$
w_V = flow velocity in the tubes $\qquad \rho_V$ = vapor density (kg/m^3)

Also, the vapor flow velocity w_V at the tube end should be at maximum 50% of the sonic velocity, w_{sonic}.

$$w_{\text{sonic}} = \sqrt{\kappa \times \frac{R}{M} \times T} \, (\text{m/s})$$

R = gas constant = 8314 $(\text{N/m}^2)(\text{m}^3)/(\text{kMol})(\text{K})$ $\qquad T$ = temperature (K)
κ = adiabatic exponent $\qquad\qquad\qquad\qquad\qquad\quad M$ = mole weight (kg/kmol)

The heat transfer coefficient in turbulent falling film can be calculated according to the following equations:

$$\text{Numrich (9.8)} \quad \alpha = 0.0055 \times \frac{\lambda}{l} \times \text{Re}^{0.44} \times \text{Pr}^{0.4} \, \left(\text{W/m}^2 \, \text{K}\right)$$

$$\text{Scholl (9.9)} \quad \alpha = 0.0054 \times \frac{\lambda}{l} \times \text{Re}^{0.38} \times \text{Pr}^{0.57} \, \left(\text{W/m}^2 \, \text{K}\right)$$

$$\text{Hadley (9.10)} \quad \alpha = 0.00454 \times \frac{\lambda}{l} \times \text{Re}^{0.44} \times \text{Pr}^{0.48} \, \left(\text{W/m}^2 \, \text{K}\right)$$

$$l = \left(\frac{v^2}{g}\right)^{1/3} \quad \text{Re} = \frac{V_U}{3600 \times v} \quad V_U = \frac{V_{\text{liq}}}{L_U} \, (\text{m}^3/\text{h m}) \quad L_U = n \times \pi \times d_i \, (\text{m})$$

Re = Reynolds number
L_U = circumferential length of the tubes (m)
V_U = liquid jet loading (m³/h m)
ν = kinematic viscosity of the product (m²/s)

Pr = Prandtl number
V_{liq} = liquid feed (m³/h)
λ = heat conductivity of the product (W/m K)

In Figure 9.16, the heat transfer coefficients calculated according to different models are shown as a function of the Reynolds number. With increasing Reynolds number or increasing liquid jet loading V_U, the heat transfer coefficient becomes greater.

Design recommendation: $V_U = 2-4$ m³/h m tube circumference.

From Figure 9.17, it follows that the heat transfer coefficient becomes better with increasing vapor velocity and higher liquid loading.

With regard to thermal damage of organic products, the residence time t_R in the evaporator is important. The residence time t_R in the falling film with the film thickness δ is determined as follows:

$$t_R = \frac{n \times d_i \times \pi \times L \times \delta}{V_{liq}/3600} \text{ (s)} \qquad \delta = 0.3035 \times Re^{0.583} \times l \text{ (m)}$$

For a uniform liquid distribution on the individual tubes of the evaporator, a liquid distributor is installed on the top tube sheet.

The distributor has on the outside an edge of ca. 100 mm and at least so many discharge pipes as evaporator tubes.

An additional predistributor is installed for large diameters.

The dimensioning of the liquid distributor is according to the equation for the vessel outlet rate V_{distr}:

$$V_{distr} = \alpha_{hole} \times n_T \times d_T^2 \times \pi/4 \times \sqrt{2 \times g \times H} \; (m^3/s)$$

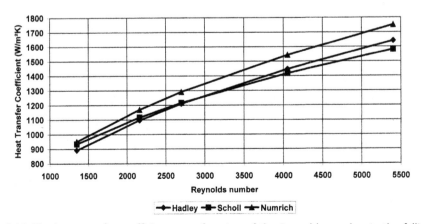

Figure 9.16 The heat transfer coefficients as a function of the Reynolds number in the falling film calculated with different models.

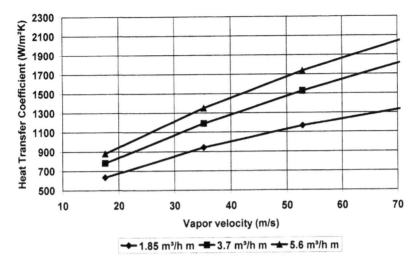

Figure 9.17 Heat transfer coefficients in a toluene evaporator as a function of vapor velocity at different liquid jet loadings.

α_{hole} = out-streaming factor = 0.6–0.8 d_T = diameter of the outflow pipes (m)
H = liquid height in the distributor (m) n_T = number of the discharge pipes

For a diameter $d_T = 15$ mm for the outflow tubes and an outflow velocity of $w_T = 0.6$ m/s in the discharge pipes, the number of the outflow tubes n_T of the distributor and the liquid height H in the distributor can be determined as follows:

$$n_T = \frac{V_{\text{liq}}}{w_T \times d_T^2 \times \pi/4 \times 3600} = \frac{V_{\text{liq}}}{0.6 \times 0.000175 \times 3600} = \frac{V_{\text{liq}}}{0.378}$$

$$H = \frac{w_T^2}{2 \times g \times \alpha_{\text{hole}}^2} = \frac{w_T^2}{2 \times 9.81 \times 0.6^2} = \frac{w_T^2}{7.06} \ (\text{m})$$

Example 17: Design of a falling film evaporator for hexadecane

Data: $P = 33$ mbar $t = 170.35\,°C$ $Q = 100$ kW $M = 226$
          $r = 75$ Wh/kg $\Delta t = 20\,°C$ $\rho_V = 0.203$ kg/m^3 $\rho_{\text{liq}} = 662$ kg/m^3
          Pr = 12.73 $\lambda = 0.094$ W/m K $\nu = 0.672$ mm^2/s   Tubes 50 × 2
          Tube length $L = 5$ m

Estimation of the required area and the required number of tubes for an estimated U value.
$U = 600$ W/m^2 K:

$$A = \frac{Q}{U \times \Delta t} = \frac{100,000}{600 \times 20} = 8.3 \ \text{m}^2$$

$$n = \frac{A}{d_o \times \pi \times L} = \frac{8.3}{0.05 \times \pi \times 5} = 11 \ \text{Rohre } 50 \times 2$$

Chosen:
16 tubes 50 × 2 with the tube length $L = 5$ m
Heat exchanger surface $A = n \times L \times \pi = 16 \times 5 \times \pi = 12.6$ m^2
Flow cross-section in the tubes $a_T = n \times d_i^2 \times \pi/4 = 16 \times 0.046^2 \times 0.785 = 0.0266$ m^2
Calculation of the *vapor flow velocity* w_V:

$$\text{Vapor density } \rho_V = \frac{M \times P + 273}{22.4 \times 1013 \times (273 + t)} = \frac{226 \times 33 \times 273}{22.4 \times 1013 \times (237 + 170)} = 0.203 \text{ kg/m}^3$$

$$\text{Vapor velocity } w_V = \frac{Q}{r \times a_T \times \rho_V \times 3600} = \frac{100,000}{75 \times 0.0266 \times 0.203 \times 3600} = 68.6 \text{ m/s}$$

Alternative calculation of the vapor flow velocity:

$$\text{Vapor rate } G_V = \frac{Q}{r} = \frac{100,000}{75} = 1333 \text{ kg/h} \qquad \text{Vapor volume } V = \frac{G_V}{\rho_V} = \frac{1333}{0.203} = 65.66 \text{ m}^3/\text{h}$$

$$\text{Vapor velocity } w_V = \frac{V}{3600 \times a_T} = \frac{65.66}{3600 \times 0.0266} = 68.6 \text{ m/s}$$

Calculation of the sonic velocity for hexadecane vapors:

$$w_{sonic} = \sqrt{\kappa \times \frac{R}{M} \times T} = \sqrt{1.0164 \times \frac{8314}{226} \times 443} = 128.7 \text{ m/s}$$

Recommendation: $w_{Vmax} \approx 0.5 \times w_{Schall} = 0.5 \times 128.7 = 64.4$ m/s.
Recommended maximum velocity of the vapors with the mass stream density.
$m_{max} = 15$ kg/m^2 s:

$$w_{Vmax} = \frac{m_{max}}{\rho_V} = \frac{15}{0.203} = 73.9 \text{ m/s}$$

The vapor velocity $w_D = 68.6$ m/s lies in the range of the allowable maximum flow velocity.
Calculation of the required *liquid feed* V_{fl} for a liquid jet loading of $V_U = 3$ m^3/h m volume:

$$V_{liq} = V_U \times L_U = V_U \times (n \times d_i \times \pi) = 3 \times 16 \times 0.046 \times \pi = 6.93 \text{ m}^3/\text{h} = 4.6 \text{ t/h}$$

Chosen: $G_{liq} = 8$ t/h $V_{liq} = 12.08$ m^3/h $V_U = 5.23$ m^3/h m

Calculation of the heat transfer coefficient for $V_U = 5.23$ m^3/h m:

$$\text{Re} = \frac{V_U}{3600 \times \nu} = \frac{5.23}{3600 \times 0.672 \times 10^{-6}} = 2161$$

$$l = \left(\frac{\nu^2}{g}\right)^{1/3} = \left(\frac{(0.672 \times 10^{-6})^2}{9.81}\right)^{1/3} = 36 \times 10^{-6}$$

$$\alpha = 0.0055 \times \frac{0.094}{36 \times 10^{-6}} \times 2161^{0.44} \times 12.73^{0.4} = 1165 \text{ W/m}^2 \text{ K}$$

Calculation of the overall heat transfer coefficient and the heat load:
Heat transfer coefficient for the heating on the shellside $\alpha_o = 6000$ W/m^2 K

Fouling rate $f_{tot} = 0.0003$ Heat conductivity of the tube wall $\lambda_{Wall} = 14$ W/m K

$$\frac{1}{U} = \frac{1}{1165 \times \dfrac{46}{50}} + \frac{1}{6000} + \frac{0.002}{14} + 0.0003 \qquad U = 648\ \text{W/m}^2\ \text{K}$$

$$Q = U \times A \times \Delta t = 648 \times 12.6 \times 20 = 163,296\ \text{W}$$

Calculation of the film thickness δ, the residence time t_R, and the falling film velocity w_{film}:

$$\delta = 0.3035 \times \text{Re}^{0.583} \times I = 0.3035 \times 2161^{0.583} \times 36 \times 10^{-6} = 0.00096\ \text{m} = 0.96\ \text{mm}$$

$$t_R = \frac{L_U \times \delta \times L}{V_{liq}/3600} = \frac{2.31 \times 0.00096 \times 5}{12.08/3600} = 3.3\ \text{s} \qquad w_{film} = \frac{L}{t_R} = \frac{5}{3.3} = 1.5\ \text{s}$$

Design of the liquid distributor for 12 m^3/h liquid feed:
Number of the required outflow tubes n_T of the distributor:

$$n_T = \frac{V_{liq}}{0.378} = \frac{12}{0.378} = 32 \quad \text{discharge pipes}$$

Liquid level H in the distributor:

$$H = \frac{w_T^2}{7.06} = \frac{0.6^2}{7.06} = 0.05\ \text{m} = 50\ \text{mm}$$

Check calculation:

$$V_{distr} = \alpha_{hole} \times n_T \times d_T^2 \times 0.785 \times \sqrt{2 \times g \times H} = 0.6 \times 32 \times 0.015^2 \times 0.785 \times \sqrt{2 \times 9.81 \times 0.05}$$
$$= 0.00336\ \text{m}^3/\text{s}$$

$$V_{liq} = V_{distr} = 3600 \times 0.00336 = 12.08\ \text{m}^3/\text{h}$$

$$w_T = \alpha_{hole} \times \sqrt{2 \times g \times H} = 0.6 \times \sqrt{2 \times 9.82 \times 0.05} = 0.6\ \text{m/s}$$

REFERENCES

[1] I.L. Mostinski, Calculations of heat transfer in boiling liquids, Teploenergetika 10 (1953) 4.
[2] M.J. McNelly, J. Imp. Coll. Chem. Eng. Soc. 7 (1953) 18.
[3] J. Starczewski, Heat transfer to nucleate boiling liquids, Brit. Chem. Eng. 10 (8) (1965).
[4] M.G. Cooper, Saturation nucleate pool boiling, J. Chem. Eng. Symp. Ser. 86 (2) (1984) 785−793.
[5] D.Q. Kern, Process Heat Transfer, McGraw-Hill, N.Y., 1950.
[6] J.W. Palen, W.M. Small, Kettle and internal reboilers, Hydrocarb. Proc. 43 (11) (1964) 199−208.
[7] Lord M. Slusser, Design parameters for condenser and reboilers, Chem. Eng. 23 (1970) 3.
[8] R.Numrich: CIT 68 (11/96), 1466 ff.
[9] S. Scholl, Verdampfung und Kondensation, in: Fluid-Verfahrenstechnik Bd. 2, Wiley-VCH-Verlag, Weinheim, 2006.
[10] M. Hadley, VDI-Fortschrittsberichte Reihe 3, Nr. 468, VDI-Verlag, 1997.

CHAPTER 10

Design of Thermosiphon Reboilers

Contents

© 2016 Elsevier Inc.
All rights reserved.

10.1 THERMAL CALCULATIONS [1–7]

10.1.1 Required circulating rate W in the thermosiphon cycle

Initially the heat load Q for the heating and evaporation for the given problem definition is determined (Figure 10.1).

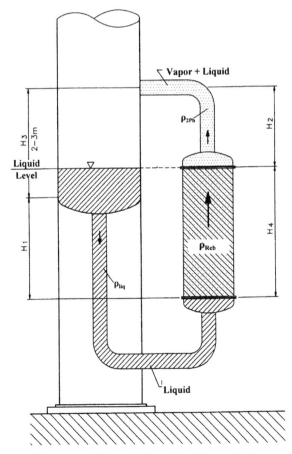

Figure 10.1 Thermosiphon circulation reboiler.

Then the required circulation rate W of the thermosiphon cycle for the chosen vapor mass fraction x is determined.

Calculation of the heat load Q for heating and evaporating:

$$Q = Q_{heat} + Q_{evap} = W \times c \times \Delta t_{heat} + W \times x \times r\,(\mathrm{W})$$

Required thermosiphon circulating rate W for the chosen evaporation rate x:

$$W = \frac{Q}{c \times \Delta t_{heat} + x \times r}\,(\mathrm{kg/h}) \quad \text{Vapour rate } G_V = x \times W\,(\mathrm{kg/h})$$

Q = heat load (W)
W = liquid inlet rate into the circulating evaporator (kg/h)
c = specific heat capacity (Wh/kg K)
Δt_{heat} = required heating up to the bubble point (K)
r = latent heat (Wh/kg)
x = vapor mass fraction in two-phase mixture (kg/kg)

Example 1: Calculation of the circulating and vapor rate in the thermosiphon evaporator

$$Q = 600\ \mathrm{kW} \qquad r = 150\ \mathrm{Wh/kg} \qquad x = 0.2 \qquad c = 0.6\ \mathrm{Wh/kg\ K} \qquad \Delta t_{heat} = 4\,^{\circ}\mathrm{C}$$

$$W = \frac{600,000}{0.6 \times 4 + 0.2 \times 150} = 18,518\ \mathrm{kg/h}$$

$$\text{Vapor rate } V = x \times W = 0.2 \times 18,518 = 3704\ \mathrm{kg/h\ vapor}$$

Cross-checking with heat balance:

$$Q = W \times c \times \Delta t + W \times x \times r = 18,518 \times 0.6 \times 4 + 18,518 \times 0.2 \times 150 = 600,000\ \mathrm{W}$$

10.1.2 Estimation of the required reboiler area A

The required area A for the required heat load Q is determined with an estimated U-value for the overall heat transfer.

$$A = \frac{Q}{U \times \Delta t}\ (\mathrm{m^2})$$

U = estimated overall heat transfer coefficient (W/m² K)

Example 2: Determination of the reboiler area and tube number

$$Q = 600\ \mathrm{kW} \qquad\qquad U = 1200\ \mathrm{W/m^2\ K} \qquad\qquad \Delta t = 20\,^{\circ}\mathrm{C}$$

$$A = \frac{Q}{U \times \Delta t} = \frac{600,000}{1200 \times 20} = 25\ \mathrm{m^2}$$

Calculation of the required number of tubes n for the reboiler area A = 25 m^2

Chosen tube length $L = 3$ m   Chosen tube dimension 25 × 2 mm   Pitch = 32 mm Δ

$$n = \frac{A}{\pi \times d_i \times L} = \frac{25}{\pi \times 0.025 \times 3} = 106 \text{ tubes } 25 \times 2$$

Chosen: $n = 127$ tubes 25 × 2 with triangular pitch 32 mm
The shell inner diameter $D_i = 384$ mm then results.

10.2 CALCULATION OF THE HEAT TRANSFER COEFFICIENT

The total heat transfer coefficient α_{Reb} is composed of the convective heat transfer coefficient α_{conv} for the convective boiling at the increased stream velocity of the two-phase mixture and α_n for the corrected nucleate boiling.

$$\alpha_{Reb} = \alpha_{conv} + \alpha_n \left(W/m^2 \text{ K}\right)$$

10.2.1 Calculation of the heat transfer coefficient α_{conv} for the convective boiling at the increased flow velocity of the two-phase mixture

Initially the α_{liq}-value for the convective heat transfer of the streaming liquid is determined.

Then the heat transfer coefficient for the proportional liquid flow α_{liq} is multiplied by the multiplicator M for the two-phase stream velocity.

$$\alpha_{conv} = M \times \alpha_{liq} \left(W/m^2 \text{ K}\right)$$

Calculation of the heat transfer coefficient α_{liq} for the streaming liquid in the tube:

Reynolds number 2300 > Re < 8000 $Nu = \left(0.037 \times Re_{liq}^{0.75} - 6.66\right) \times Pr^{0.42}$

Reynolds number > 8000 $Nu = 0.023 \times Re_{liq}^{0.8} \times Pr^{0.33}$

$$\alpha = \frac{Nu \times \lambda}{d_i} \quad Re_{liq} = \frac{w_{liq} \times d_i}{\nu_{liq}}$$

Re_{liq} = Reynolds number
Pr = Prandtl number
Nu = Nusselt number
λ = heat conductivity (W/m K)
d_i = tube inner diameter
ν_{liq} = kinematic viscosity (m^2/s)
w_{liq} = flow velocity of the liquid (m/s)

The multiplicator M for the conversion of α_{liq} to α_{conv} at the two-phase streaming is calculated as follows:

$$M = 2.5 \times \left(\frac{1}{X_{\text{tt}}}\right)^{0.7}$$

$$X_{\text{tt}} = \left(\frac{1-x}{x}\right)^{0.9} \times \left(\frac{\rho_V}{\rho_{\text{liq}}}\right)^{0.5} \times \left(\frac{\eta_{\text{liq}}}{\eta_V}\right)^{0.1} = \left(\frac{G_{\text{liq}}}{G_V}\right)^{0.9} \times \left(\frac{\rho_V}{\rho_{\text{liq}}}\right)^{0.5} \times \left(\frac{\eta_{\text{liq}}}{\eta_V}\right)^{0.1}$$

X_{tt} = Lockhart–Martinelli parameter for turbulent streaming
x = vapor weight composition of the two-phase streaming
ρ_{liq} = liquid density (kg/m^3)
ρ_V = vapor density (kg/m^3)
η_{liq} = dynamic liquid viscosity (mPa s)
η_V = dynamic vapor viscosity (mPa s)
G_{liq} = liquid rate (kg/h)
G_V = vapor rate (kg/h)

Example 3: Calculation of the heat transfer coefficient for corrected convective boiling in a vertical thermosiphon reboiler for toluene

Circulating rate $W = 40$ t/h $x = 0.4 = 40\%$ evaporation at the outlet = 16 t/h vapor

The vapor mass fraction increases from $x = 0$ at the reboiler inlet to $x = 0.4$ at the reboiler outlet.
The calculation is made with the average vapor mass fraction x_a and the average vapor rate G_{Va} and the average liquid rate G_{liqa}
Average vapor mass fraction over the reboiler tube length $x_a = 0.2$

Average liquid rate $G_{\text{liqa}} = 32$ t/h $= 41$ m^3/h   Average vapor rate $G_{Va} = 8$ t/h
$\rho_V = 3$ kg/m^3 $\rho_{\text{liq}} = 781$ kg/m^3 $\eta_V = 0.009$ mPa s $\eta_{\text{liq}} = 0.244$ mPa s
$\lambda_{\text{liq}} = 0.116$ W/m K $Pr = 4.12$ $d_i = 21$ mm $\nu_{\text{liq}} = 0.3073$ mm^2/s

Calculation of the flow cross-section a_t of the tubes and the flow velocity w_{liq}:

$$a_t = n \times d_i^2 \times \frac{\pi}{4} = 120 \times 0.021^2 \times 0.785 = 0.04154 \text{ m}^2 \qquad w_{\text{liq}} = \frac{41}{3600 \times 0.04154} = 0.274 \text{ m/s}$$

Calculation of α_{liq}:

$$Re_{\text{liq}} = \frac{w_{\text{liq}} \times d_i}{\nu} = \frac{0.274 \times 0.021}{0.3073 \times 10^{-6}} = 18,723$$

$$Nu = 0.023 \times Re^{0.8} \times Pr^{0.33} = 0.023 \times 18,723^{0.8} \times 4.12^{0.33} = 96.06$$

$$\alpha_{\text{liq}} = \frac{Nu \times \lambda}{d_i} = \frac{96.06 \times 0.116}{0.021} = 530 \text{ W/m}^2 \text{ K}$$

Calculation of M and α_{covn}

$$X_{tt} = \left(\frac{G_{liq}}{G_V}\right)^{0.9} \times \left(\frac{\rho_V}{\rho_{liq}}\right)^{0.5} \times \left(\frac{\eta_{liq}}{\eta_V}\right)^{0.1} = \left(\frac{32}{8}\right)^{0.9} \times \left(\frac{3}{781}\right)^{0.5} \times \left(\frac{0.244}{0.009}\right)^{0.1} = 0.3$$

$$M = 2.5 \times \left(\frac{1}{X_{tt}}\right)^{0.7} = 2.5 \times \left(\frac{1}{0.3}\right)^{0.7} = 5.8$$

$$\alpha_{conv} = M \times \alpha_{liq} = 5.8 \times 530 = 3074 \text{ W/m}^2 \text{ K}$$

10.2.2 Calculation of the heat transfer coefficient α_N for the nucleate boiling with the correction factor S for confined nucleate boiling

Initially the heat transfer coefficient α_{NB} for the nucleate boiling is determined.

This value will be corrected with the correction factor S according to Chen for the confined nucleate boiling at the two-phase streaming.

$$\alpha_{NB} = 0.00626 \times P_c^{0.69} \times q^{0.7} \left(\text{W/m}^2 \text{ K}\right) \quad \alpha_n = S \times \alpha_{NB} \left(\text{W/m}^2 \text{ K}\right)$$

$$S = \frac{1}{1 + \left(2.53 \times 10^{-6} \times \text{Re}_{2Ph}^{1.17}\right)} \quad \text{Re}_{2Ph} = \text{Re}_{liq} \times M^{1.25}$$

q = heat flux density (W/m^2)
P_c = critical pressure (kPa)
S = correction factor for the confined nucleate boiling according to Chen
Re_{2Ph} = Reynolds number at two-phase streaming according to Chen

---

Example 4: Heat transfer coefficient for corrected nucleate boiling

$P_c = 4100$ kPa $q = 25,000$ W/m^2 Re $= 18,723$ $M = 5.8$

$$\alpha_{NB} = 0.00626 \times 4100^{0.69} \times 25,000^{0.7} = 2333 \text{ W/m}^2 \text{ K}$$

$$\text{Re}_{2Ph} = 18,723 \times 5.8^{1.25} = 168,523 \quad S = \frac{1}{1 + \left(2.53 \times 10^{-6} \times 168,523^{1.17}\right)} = 0.23$$

$$\alpha_n = 2333 \times 0.23 = 536 \text{ W/m}^2 \text{ K}$$

The total heat transfer coefficient for the evaporation in the vertical thermosiphon evaporator is the sum of the heat transfer coefficient for the convective heat transfer of the two-phase streaming (Example 3) and the corrected heat transfer coefficient for the prevented nucleate boiling by the streaming (Example 4):

$$\alpha_{Reb} = \alpha_{conv} + \alpha_N = 3074 + 536 = 3610 \text{ W/m}^2 \text{ K}$$

10.2.3 Calculations for a vertical thermosiphon evaporator

The calculation is performed in the following steps:
- determination of the flow cross–section in the tubes a_t
- determination of the flow velocity w_{liq} for the liquid rate
- calculation of the Reynolds number Re_{liq}
- determination of the heat transfer coefficient α_{liq} for the liquid flowing in the tubes

---

Example 5: Thermal design of a steam-heated vertical thermosiphon evaporator type AEM with evaporation in the tubes

Data: average liquid rate $V = 85$ m³/h

Reboiler DN 700 with 394 tubes 25 × 2 $L = 2.5$ m, flow   Cross-section $a_t = 0.136$ m²
$Pr = 4.33$ $\lambda = 0.117$ W/m K $N = 0.318$ mm²/s

Calculation of the heat transfer coefficient α_{conv} for the convective boiling:

$$w_{liq} = \frac{85}{3600 \times 0.136} = 0.173 \text{ m/s} \quad Re_{liq} = \frac{0.173 \times 0.021}{0.318 \times 10^{-6}} = 11,431$$

$$\alpha_{liq} = 0.023 \times Re^{0.8} \times Pr^{0.33} \times \frac{\lambda}{d_i} = 0.023 \times 11,431^{0.8} \times 4.33^{0.33} \times \frac{0.117}{0.021} = 366 \text{ W/m}^2 \text{ K}$$

Two-phase multiplicator $M = 4$

$$\alpha_{conv} = 4 \times 366 = 1466 \text{ W/m}^2 \text{ K}$$

Calculation of the heat transfer coefficient for the nucleate boiling:

$$P_c = 4100 \text{ kPa} \qquad q = 20,000 \text{ W/m}^2$$
$$\alpha_{NB} = 0.00626 \times P_c^{0.69} \times q^{0.7} = 0.00626 \times 4100^{0.69} \times 20,000^{0.7} = 1996 \text{ W/m}^2 \text{ K}$$
$$Re_{2Ph} = 11,431 \times 4^{1.25} = 64,663 \quad S = 0.48$$
$$\alpha_n = 0.48 \times 1996 = 958 \text{ W/m}^2 \text{ K}$$

Total heat transfer coefficient $\alpha_{Reb} = \alpha_{conv} + \alpha_N = 1466 + 958 = 2424$ W/m² K
Calculation of the overall heat transfer coefficient U:

$$f_{tot} = 0.0003 \qquad s = 2 \text{ mm} \qquad \lambda_{Wall} = 14 \text{ W/m K}$$
Tubeside $\alpha_{io} = 2424 \times \frac{21}{25} = 2036 \text{ W/m}^2 \text{ K}$ Shell side $\alpha_o = 8000 \text{ W/m}^2 \text{ K}$

$$\frac{1}{U} = \frac{1}{2036} + \frac{1}{8000} + \frac{0.002}{14} + 0.0003 \qquad U = 944 \text{ W/m}^2 \text{ K}$$

Calculation of the heat load Q with logarithmic mean temperature difference (LMTD) = 20 °C:

$$A = 394 \times 0.025 \times \pi \times 2.5 = 77 \text{ m}^2$$
$$Q = U \times A \times \text{LMTD} = 944 \times 77 \times 20 = 1.454 \text{ MW}$$
$$q = U \times \text{LMTD} = 944 \times 20 = 18,880 \text{ W/m}^2$$

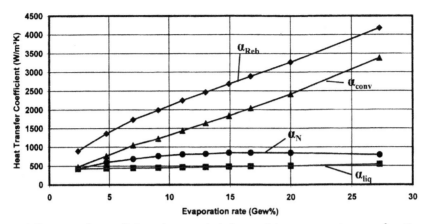

Figure 10.2 Heat transfer coefficients in a vertical thermosiphon evaporator as a function of the evaporation rate.

From Figure 10.2 it follows that the heat transfer coefficient α_{Reb} increases with increasing vaporization because the heat transfer coefficient α_{kon} for the convective boiling of the two-phase mixture strongly increases.

Dependences on the vapor composition x

- The two-phase density decreases with increasing vaporization and the flow velocity of the two-phase mixture increases. The higher flow velocity improves the heat transfer coefficient for the convective boiling.
- The overall heat transfer coefficient increases at higher evaporation rates because the α-value for the convective boiling increases with increasing vapor rate.
- The effective temperature difference LMTD increases at larger evaporation rates because due to the lower density in the reboiler at higher vapor compositions the boiling pressure and hence the boiling temperature are reduced.

10.2.4 Calculations for a horizontal thermosiphon evaporator with cross flow between the tubes from the bottom to the top [8,9]

Calculation steps:

- determination of the flow cross-section for the cross streams around the tubes a_{cross}
- determination of the flow velocity w_{cross} for the liquid rate
- calculation of the Reynolds number Re_{liq}
- determination of the heat transfer coefficient for the streaming liquid α_{liq} across the tubes in the square arrangement.

$$\alpha_{liq} = 0.158 \times Re^{0.6} \times Pr^{0.33} \times \frac{\lambda}{d_o} \left(W/m^2\ K\right) \quad \alpha_{conv} = M \times \alpha_{liq}\left(W/m^2\ K\right)$$

$$\alpha_{Reb} = \alpha_{conv} + \alpha_n\left(W/m^2\ K\right)$$

Example 6: Thermal design of a horizontal thermosiphon evaporator type AJL or AHL with shell-side cross flow evaporation

Tube bundle DN 450 with 132 tubes 25 × 2, 6 m long, 32 mm quadratic pitch

Physical data of Example 5

Calculation of the heat transfer coefficient α_{conv} for the convective boiling:

$$\text{Number of tubes in cross flow} \quad n_{cross} = \frac{D_i}{T} = \frac{450}{32} = 14$$

$$a_{cross} = D_i - n_{cross} \times d_o = 0.45 - 14 \times 0.025 = 0.1 \, m^2 \text{ cross flow area}$$

$$w_{cross} = \frac{85}{3600 \times 0.1} = 0.236 \, m/s \quad Re = \frac{0.236 \times 0.025}{0.318 \times 10^{-6}} = 18,562$$

$$\alpha_{liq} = 0.158 \times Re^{0.6} \times Pr^{0.33} \times \frac{\lambda}{d_o} = 0.158 \times 18,562^{0.6} \times 4.33^{0.33} \times \frac{0.117}{0.025} = 436.6 \, W/m^2 \, K$$

$$\text{Two-phase multiplicator} = 4 \qquad \alpha_{conv} = 4 \times 436.6 = 1746 \, W/m^2 \, K$$

Calculation of the heat transfer coefficient α_N for the nucleate boiling:

$$P_c = 4100 \, kPa \qquad\qquad q = 20,000 \, W/m^2$$

$$\alpha_{NB} = 0.00626 \times 4100^{0.69} \times 25,000^{0.7} = 1996 \, W/m^2 \, K$$

$$Re_{2Ph} = 18,562 \times 4^{1.25} = 105,002 \quad S = 0.345 \quad \alpha_n = 0.345 \times 1996 = 688 \, W/m^2 \, K$$

Total heat transfer coefficient $\alpha_{Reb} = \alpha_{conv} + \alpha_N = 1746 + 688 = 2434 \, W/m^2 \, K$

Calculation of the overall heat transfer coefficient U:

$$\text{Tubeside} \quad \alpha_{io} = 8000 \times \frac{21}{25} = 6720 \, W/m^2 \, K \quad \text{Shell side} \quad \alpha_{Reb} = 2434 \, W/m^2 \, K$$

$$\frac{1}{U} = \frac{1}{2434} + \frac{1}{6720} + \frac{0.002}{14} + 0.0003 \qquad U = 998 \, W/m^2 \, K$$

Calculation of the heat load Q for LMTD = 20 °C:

$$\text{Reboiler area} \quad A = 132 \times 0.025 \times \pi \times 6 = 62 \, m^2$$

$$Q = U \times A \times LMTD = 998 \times 62 \times 20 = 1.237 \, MW$$

$$\text{Heat flux density} \, q = U \times LMTD = 998 \times 20 = 19,960 \, W/m^2$$

10.2.5 Calculations for a horizontal thermosiphon evaporator with longitudinal flow around the tubes [8,9]

Calculation steps:

- determination of the flow cross-section for the longitudinal streaming around the tubes a_{shell}
- determination of the flow velocity w_{liq} for the liquid rate
- determination of the hydraulic diameter d_H
- calculation of the Reynolds number
- determination of the heat transfer coefficient for the liquid α_{liq} flowing shell side

$$a_{shell} = \frac{\pi}{4} \times \left(D_i^2 - n \times d_o^2\right)\left(m^2\right) \qquad d_H = \frac{D_i^2 - n \times d_o^2}{n \times d_o}(m)$$

$$\alpha_{liq} = 0.023 \times Re^{0.8} \times Pr^{0.33} \times \frac{\lambda}{d_o}\left(W/m^2\,K\right) \quad \text{for } Re > 8000$$

$$\alpha_{conv} = M \times \alpha_{liq}\left(W/m^2\,K\right) \qquad \alpha_{Reb} = \alpha_{conv} + \alpha_n\left(W/m^2\,K\right)$$

Example 7: Thermal design of a horizontal thermosiphon evaporator type AEL with longitudinal flow through the shell without baffles

DN 450 with 132 tubes 25 × 2, 6 m long, 32 mm quadratic pitch

Calculation of the heat transfer coefficient for the convective boiling on the shell side:

$$a_{shell} = \frac{\pi}{4} \times \left(D_i^2 - n \times d_o^2\right) = 0.785 \times \left(0.45^2 - 132 \times 0.025^2\right) = 0.094\ m^2$$

$$w_{liq} = \frac{85}{3600 \times 0.094} = 0.25\ m/s$$

$$d_H = \frac{D_i^2 - n \times d_o^2}{n \times d_o} = \frac{0.45^2 - 132 \times 0.025^2}{132 \times 0.025} = 0.036\ m$$

$$Re = \frac{w_{liq} \times d_H}{v} = \frac{0.25 \times 0.036}{0.318 \times 10^{-6}} = 28,301$$

$$\alpha_{liq} = 0.023 \times 28,301^{0.8} \times 4.33^{0.33} \times \frac{0.117}{0.036} = 441.7\ W/m^2\,K$$

Multiplicator for two-phase flow $M = 4$

$$\alpha_{conv} = 4 \times 441.7 = 1767\ W/m^2\,K$$

Calculation of the heat transfer coefficient for the nucleate boiling:

$$P_c = 4100\ kPa \qquad\qquad q = 20,000\ W/m^2$$
$$\alpha_{NB} = 0.00626 \times 4100^{0.69} \times 20,000^{0.7} = 1996\ W/m^2\,K$$
$$Re_{2Ph} = 28,301 \times 4^{1.25} = 160,095 \quad S = 0.243 \quad \alpha_n = 0.243 \times 1996 = 486\ W/m^2\,K$$

Total heat transfer coefficient $\alpha_{Reb} = \alpha_{conv} + \alpha_N = 1767 + 486 = 2253\ W/m^2\,K$

Calculation of the overall heat transfer coefficient U:

$$\text{Tubeside} \quad \alpha_{io} = 6720\ W/m^2\,K \qquad \text{Shell side} \quad \alpha_{Reb} = 2253\ W/m^2\,K$$

$$\frac{1}{U} = \frac{1}{2253} + \frac{1}{6720} + \frac{0.002}{14} + 0.0003 \quad U = 966\ W/m^2\,K$$

Calculation of the heat load Q for LMTD = 20 °C:

$$\text{Reboiler area } A = 132 \times 0.025 \times \pi \times 6 = 62\ m^2$$
$$Q = U \times A \times LMTD = 966 \times 62 \times 20 = 1.2\ MW$$

Heat flux density $q = U \times LMTD = 960 \times 22 = 19,320\ W/m^2$

10.3 CALCULATION OF THE TWO-PHASE DENSITY AND THE AVERAGE DENSITY IN THE REBOILER

10.3.1 Calculation of the two-phase density ρ_{2Ph} of a vapor—liquid mixture

$$\rho_{2Ph} = \frac{1}{\frac{x}{\rho_V} + \frac{1-x}{\rho_{liq}}} \left(kg/m^3\right)$$

x = vapor fraction in the mixture
ρ_V = vapor density (kg/m^3)
ρ_{liq} = liquid density (kg/m^3)

10.3.2 Calculation of the average density ρ_{Reb} in the reboiler from the density of the liquid ρ_{liq} and the density of the two-phase mixture ρ_{2Ph}

$$\rho_{Reb} = \frac{1}{\frac{1}{\rho_{2Ph}} - \frac{1}{\rho_{liq}}} \times \ln\left(\frac{\left(\frac{1}{\rho_{2Ph}} - \frac{1}{\rho_{liq}}\right)}{\frac{1}{\rho_{liq}}}\right)$$

Calculation of the average density ρ_{Reb} in the reboiler from the vapor density ρ_V and the liquid density ρ_{liq} and the vapor fraction x:

$$\rho_{Reb} = \frac{1}{x \times \left(\frac{1}{\rho_V} - \frac{1}{\rho_{liq}}\right)} \times \ln\left(\frac{x \times \left(\frac{1}{\rho_V} - \frac{1}{\rho_{liq}}\right)}{\frac{1}{\rho_{liq}}}\right)$$

Example 8: Calculation of the two-phase density ρ_{2Ph} and the average density ρ_{Reb} in the reboiler (Figure 10.3)

$x = 0.136$ Vapor density $\rho_V = 0.68 \ kg/m^3$ Liquid density $\rho_{liq} = 750 \ kg/m^3$

Calculation of the two-phase density of the mixture:

$$\rho_{2Ph} = \frac{1}{\frac{x}{\rho_V} + \frac{1-x}{\rho_{liq}}} = \frac{1}{\frac{0.136}{0.68} + \frac{0.864}{750}} = 4.97 \ kg/m^3$$

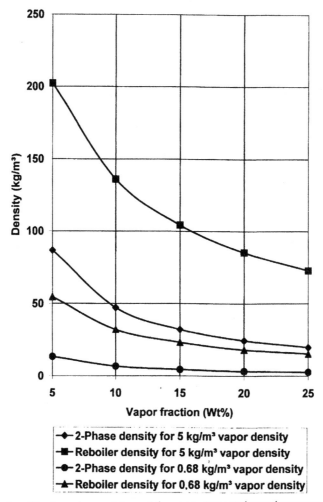

Figure 10.3 Density of the two-phase mixture at the evaporator outlet and average reboiler density in the evaporator as a function of the vapor rate in the two-phase mixture for two different vapor densities.

Calculation of the average density in the reboiler:

$$\rho_{Reb} = \frac{1}{\frac{1}{4.97} - \frac{1}{750}} \times \ln\left(\frac{\frac{1}{4.97} - \frac{1}{750}}{\frac{1}{750}}\right) = 25.07 \text{ kg/m}^3$$

$$\rho_{Reb} = \frac{1}{0.136 \times \left(\frac{1}{0.68} - \frac{1}{750}\right)} \times \ln\left(\frac{0.136 \times \left(\frac{1}{0.68} - \frac{1}{750}\right)}{\frac{1}{750}}\right) = 25.07 \text{ kg/m}^3$$

10.4 FLOW VELOCITY W_{REB} IN THE REBOILER

$$w_{Reb} = \frac{G}{\rho \times a \times 3600} \, (m/s)$$

a = flow cross-section (m^2)

G = mass flow rate through the evaporator (kg/h)

ρ = density of the two-phase mixture (kg/m^3) or the liquid or the average density in the reboiler

---

Example 9: Calculation of the two-phase flow velocity in the vertical reboiler

Reboiler inlet: 20 t/h
Density of the liquid: 750 kg/m³
Flow cross-section = 0.06474 m² for 187 tubes 25 × 2
Flow velocity for the incoming liquid: w_{ein} = 0.114 m/s
Outlet from the reboiler = 20 t/h with 20 weight% vapor
Vapor density ρ_G = 5 kg/m³
Density of the two-phase mixture ρ_{2Ph} = 24.35 kg/m³ at the evaporator outlet
Flow velocity for the outflowing two-phase mixture: w_{out} = 3.52 m/s
Logarithmic averaged velocity w_{av} in the 3-m-long evaporator tubes:

$$w_{av} = \frac{w_{out} - w_{in}}{\ln \frac{w_{out}}{w_{in}}} = \frac{3.52 - 0.114}{\ln \frac{3.52}{0.114}} = 1 \, m/s$$

Average density ρ_{Reb} in the reboiler:

$$\rho_{Reb} = \frac{1}{\frac{1}{24.35} - \frac{1}{750}} \times \ln \left(\frac{\frac{1}{24.35} - \frac{1}{750}}{\frac{1}{750}} \right) = 85.43 \, kg/m^3$$

Hence the average flow velocity in the reboiler tubes

$$w_{av} = \frac{20,000}{85.43 \times 0.06474 \times 3600} = 1 \, m/s$$

Alternative calculation for the vapor density ρ_V = 0.68 kg/m³

Two-phase density at the evaporator outlet for x = 0.2: ρ_{2Ph} = 3.388 kg/m³

Flow velocity at the evaporator outlet: 25.32 m/s
Average flow velocity in the reboiler: w_a = 4.67 m/s
Average density in the reboiler ρ_{Reb} = 18.36 kg/m³

$$w_{av} = \frac{20,000}{18.36 \times 3600 \times 0.06474} = 4.67 \, m/s$$

10.5 DETERMINATION OF THE REQUIRED HEIGHT *H*1 FOR THE THERMOSIPHON CIRCULATION OR THE MAXIMUM ALLOWABLE PRESSURE LOSS Δ*P*_{MAX} IN THERMOSIPHON CIRCULATION [10]

10.5.1 Calculation of the required elevation height *H*1 for overcoming the pressure loss Δ*P* in the thermosiphon circulation with the safety factor *S*

$$H1 = \frac{S \times \Delta P + g \times \left(H2 \times \rho_{2\text{ph}} + H4 \times \rho_{\text{Reb}}\right)}{g \times \rho_{\text{liq}}} \text{(m liquid height)}$$

Calculation of the required driving pressure Δ*P*_{req} *for thermosiphon circulation:*

$$\Delta P_{\text{req}} = \rho_{\text{liq}} \times g \times H1 = S \times \Delta P + g \times (\rho_{2\text{Ph}} \times H2 + \rho_{\text{Reb}} \times H4)(\text{Pa})$$

10.5.2 Calculation of the maximum allowable pressure loss Δ*P*_{allow} in the thermosiphon circulation at given heights *H*1, *H*2, and *H*4

$$\Delta P_{\text{allow}} = g \times \left(H1 \times \rho_{\text{liq}} - H2 \times \rho_{2\text{Ph}} - H4 \times \rho_{\text{Reb}}\right) \text{Pa}$$

ΔP = pressure loss in the thermosiphon circulation (Pa)
$H1$ = driving liquid height (m) (see Figure 10.4)
ρ_{liq} = liquid density (kg/m^3)
$H2$ = height of the two-phase mixture in the riser (m)
$\rho_{2\text{Ph}}$ = two-phase density (kg/m^3)
$H4$ = height of the reboiler (m)
ρ_{Reb} = average density in the reboiler (kg/m^3)
S = safety factor = 2

Example 10: Calculation of the required height *H*1 for thermosiphon circulation

$\Delta P = 69.2 \text{ mbar} = 6920 \text{ Pa}$ $\rho_{\text{liq}} = 750 \text{ kg/m}^3$
$H2 = 2 \text{ m}$ $\rho_{2\text{Ph}} = 4.97 \text{ kg/m}^3$ $H4 = 3 \text{ m}$ $\rho_{\text{Reb}} = 25.07 \text{ kg/m}^3$

Safety factor $S = 2$
Calculation of the required elevation height *H*1 for a pressure loss of $\Delta P = 69.2$ mbar:

$$H1 = \frac{2 \times 6920 + 9.81 \times (2 \times 4.97 + 3 \times 25.07)}{9.81 \times 750} = 1.99 \text{ m elevation height with } S = 2$$

Required driving pressure for circulation:

$$\Delta P_{\text{req}} = \rho_{\text{liq}} \times g \times H1 = 750 \times 9.81 \times 1.99 = 14,641 \text{ Pa}$$

Maximum allowable pressure loss at $H1 = 1.99$ m:

$$\Delta P_{\text{allow}} = 9.81 \times (1.99 \times 750 - 2 \times 4.97 - 3 \times 25.07) = 13,840 \text{ Pa} = 2 \times \Delta P$$

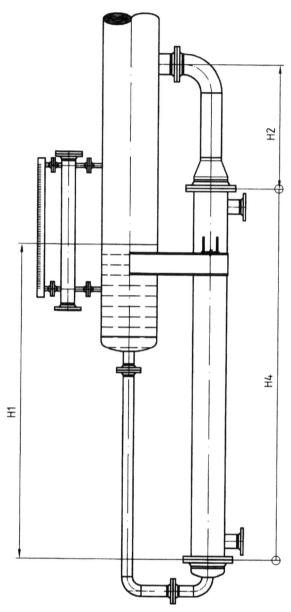

Figure 10.4 Thermosiphon evaporator with the heights $H1 + H2 + H4$ for the determination of the driving pressure for the thermosiphon circulation.

Notice:

If a part of the tube length is required for heating up the liquid to the boiling temperature this length must be considered when determining $H1_{netto}$.

$$H1_{netto} = H1 - H_{heat}$$

$$H_{heat} = \text{required heating length (m)}$$

Example 11: Calculation of the available driving pressure Δp_{avail} for the thermosiphon circulation

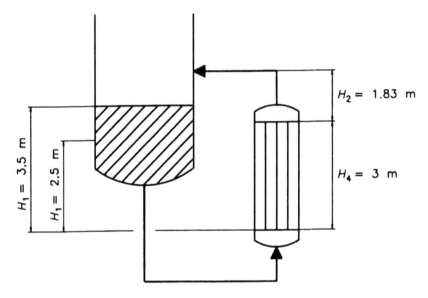

The calculation is performed for the heights $H1 = 2.5$ m and $H1 = 3.5$ m.

$H2 = 1.83$ m $\rho_2 = 24.3$ kg/m^3

$H4 = 3$ m $\rho_3 = 85.3$ kg/m^3

$\rho_1 = 750$ kg/m^3

$$\Delta P_{avail} = g \times \left(H1 \times \rho_{liq} - H2 \times \rho_{2Ph} - H4 \times \rho_{Reb}\right)$$

$H1 = 2.5$ m

$$\Delta P_{avail} = 9.81 \times (2.5 \times 750 - 1.83 \times 24.3 - 3 \times 85.3) = 15,447 \text{ Pa} = 154 \text{ mbar}$$

$H1 = 3.5$ m

$$\Delta P_{avail} = 9.81 \times (3.5 \times 750 - 1.83 \times 24.3 - 3 \times 85.3) = 22,805 \text{ Pa} = 228 \text{ mbar}$$

10.5.3 Calculation of the thermosiphon reboiler feed height H1 in practice

10.5.3.1 Vertical circulating reboiler (see Figure 10.5)

Reference points for $H1$: weld seam torospherical head/column frame

Calculation of the minimum feed height $H1_{min}$ with the safety factor 2:

$$H1_{min} = \frac{2 \times \Delta P + H2 \times \rho_{2Ph} + H4 \times \rho_{Reb}}{\rho_{liq}}$$

$$= \frac{2 \times \Delta P + \rho_{2Ph} \times (H3 - H4) + H4 \times \rho_{Reb}}{\rho_{liq} - \rho_{2Ph}}(m)$$

ΔP = pressure loss in the thermosiphon circuit (mm liquid height)

$$H2 = H1 + H3 - H4(m)$$

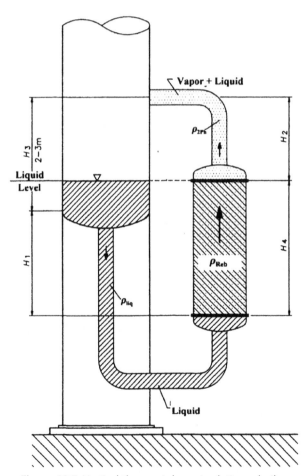

Figure 10.5 Vertical thermosiphon circulation reboiler.

Example 12: Calculation of the minimum feed height $H1$ for a vertical circulation reboiler

$$\Delta P = 300 \text{ mm liquid height} = 0.3 \times 9.81 \times 750 = 2207 \text{ Pa}$$

$$H3 = 2.5 \text{ m} \qquad\qquad H4 = 3 \text{ m}$$

$$\rho_{liq} = 750 \text{ kg/m}^3 \qquad\quad \rho_{2Ph} = 24.3 \text{ kg/m}^3 \qquad\quad \rho_{Reb} = 85.3 \text{ kg/m}^3$$

$$H1 = \frac{2 \times 300 + 24.3 \times (2.5 - 3) + 85.3 \times 3}{750 - 24.3} = 1.16 \text{ m}$$

$$H2 = H1 + H3 - H4 = 1.16 + 2.5 - 3 = 0.66 \text{ m}$$

$$H1 = \frac{2 \times 300 + 24.3 \times 0.66 + 85.3 \times 3}{750} = 1.16 \text{ m}$$

Cross-check:

$$\Delta H_{allow} = H1 \times \rho_{liq} - H2 \times \rho_{2Ph} - H4 \times \rho_{Reb} = 1.16 \times 750 - 0.66 \times 24.3 - 3 \times 85.3$$

$$= 600 \text{ mm liquid height}$$

$$\Delta P_{allow} = \frac{\Delta H_{allow}}{1000} \times g \times \rho_{liq} = 0.6 \times 9.81 \times 750 = 4414 \text{ Pa} = 2 \times \Delta P$$

10.5.3.2 Horizontal circulation evaporator (see Figure 10.6)

Reference points for $H1$: weld seam torospherical head/column frame

Calculation of $H1$ with the safety factor 2:

$$H1 = \frac{2 \times \Delta P + H3 \times \rho_{2Ph}}{\rho_{liq} - \rho_{2Ph}} = \frac{2 \times \Delta P + H2 \times \rho_{2Ph}}{\rho_{liq}} \text{ (m)} \quad H2 = H1 + H3 \text{(m)}$$

Example 13: Calculation of the minimum feed height $H1$ for a horizontal circulation reboiler

$$\Delta P = 500 \text{ mm liquid height} = 3679 \text{ Pa} \qquad H3 = 2.5 \text{ m}$$

$$\rho_{liq} = 750 \text{ kg/m}^3 \qquad\qquad\qquad \rho_{2Ph} = 24.3 \text{ kg/m}^3$$

$$H1 = \frac{2 \times 500 + 24.3 \times 2.5}{750 - 24.3} = 1.46 \text{ m} \quad H2 = H1 + H3 = 1.46 + 2.5 = 3.96 \text{ m}$$

$$H1 = \frac{2 \times 500 + 24.3 \times 3.96}{750} = 1.46 \text{ m}$$

Cross-check:

$$\Delta H = \rho_{liq} \times H1 - \rho_{2Ph} \times H2 = 750 \times 1.46 - 24.3 \times 3.96 = 1000 \text{ mm liquid height}$$

$$\Delta P_{allow} = \frac{\Delta H}{1000} \times g \times \rho_{liq} = 1 \times 9.81 \times 750 = 7358 \text{ Pa} = 2 \times \Delta P$$

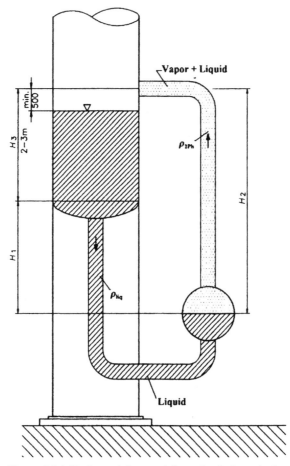

Figure 10.6 Horizontal thermosiphon circulation reboiler.

10.5.3.3 Vertical once-through evaporator (see Figure 10.7)

Reference point for $H1$: height of the liquid draw support nozzle at the column

With the once-through evaporator, more driving height $H1$ mostly exists than necessary.

Calculation of $H1$ with safety factor $S = 2$:

$$H1 = \frac{S \times \Delta P - \rho_{2Ph} \times (H4 + \Delta H) + \rho_{Reb} \times H4}{\rho_{liq} - \rho_{2Ph}}$$

$$= \frac{S \times \Delta P + \rho_{2Ph} \times H2 + \rho_{Reb} \times H4}{\rho_{liq}} \, (m)$$

ΔP = pressure loss in the thermosiphon circuit (mm liquid height)

The distance between the draw nozzle and the vapor return nozzle is normally $\Delta H = 1$ m.

$$H2 = H1 - H4 - 1 \, (m) \quad H1 = H2 + H4 + 1$$

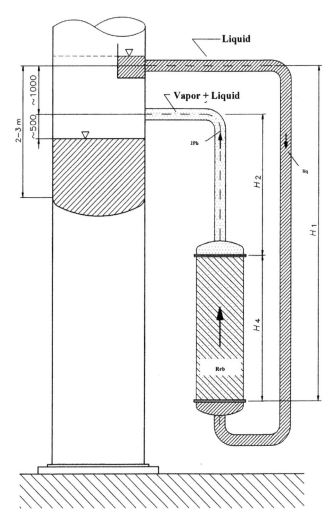

Figure 10.7 Vertical thermosiphon once-through reboiler.

Example 14: Calculation of the minimum feed height $H1$ for a vertical once-through reboiler

$\Delta P = 2000$ mm   liquid height $= 14{,}715$ Pa $H4 = 3$ m $\Delta H = 1$ m
$\rho_{liq} = 750$ kg/m^3 $\rho_{2Ph} = 24.3$ kg/m^3 $\rho_{Reb} = 85.3$ kg/m^3

$$H1 = \frac{2 \times 2000 - 24.3 \times (3+1) + 85.3 \times 3}{750 - 24.3} = 5.73 \text{ m} \quad H2 = 5.73 - 3 - 1 = 1.73 \text{ m}$$

$$H1 = \frac{2 \times 2000 + 24.3 \times 1.73 + 85.3 \times 3}{750} = 5.73 \text{ m}$$

Cross-check:

$$\Delta H_{allow} = \rho_{liq} \times H1 - \rho_{2Ph} \times H2 - \rho_{Reb} \times H4 = 750 \times 5.73 - 1.73 \times 24.3 - 3 \times 86.3$$
$$= 4000 \text{ mm liquid height}$$
$$\Delta P_{allow} = 4 \times 9.81 \times 750 = 29,430 \text{ Pa} = 2 \times \Delta P$$

10.5.3.4 Horizontal once-through evaporator (see Figure 10.8)

Reference point for $H1$: height of the liquid draw nozzle at the column

Constructive requirement: $H2 = H1 - 1$ m

In most cases, with the once-through evaporator, the given driving height $H1$ is much more than required.

Calculation of the elevation height $H1$ with the safety factor 2:

$$H1 = \frac{2 \times \Delta P + \rho_{2Ph} \times H2}{\rho_{liq}} = \frac{2 \times \Delta P - \rho_{2Ph} \times 1}{\rho_{liq} - \rho_{2Ph}} \text{ (m)}$$

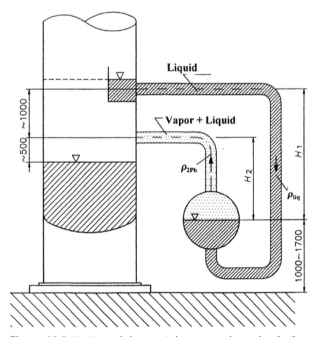

Figure 10.8 Horizontal thermosiphon once-through reboiler.

Example 15: Calculation of the minimum feed height $H1$ for a horizontal once-through reboiler

$\Delta P = 1000$ mm   liquid height = 7358 Pa $H1 - H2 = 1$ m

$\rho_{liq} = 750$ kg/m^3 $\rho_{2Ph} = 24.3$ kg/m^3 $\rho_{Reb} = 85.3$ kg/m

$$H1 = \frac{2 \times 1000 - 24.3 \times 1}{750 - 24.3} = 2.72 \text{ m} \qquad H2 = H1 - 1 = 2.72 - 1 = 1.72$$

$$H1 = \frac{2 \times 1000 - 24.3 \times 1.72}{750} = 2.72 \text{ m}$$

Cross-check:

$$\Delta H_{allow} = \rho_{liq} \times H1 - \rho_{2Ph} \times H2 = 750 \times 2.72 - 24.3 \times 1.72 = 2000 \text{ mm liquid height}$$

$$\Delta P_{allow} = \frac{\Delta H_{allow}}{1000} \times g \times \rho_{liq} = \frac{2000}{1000} \times 9.81 \times 750 = 14,715 \text{ Pa} = 2 \times \Delta P$$

10.5.4 Effects of the elevation height $H1$ on the circulating rate, the pressure loss, the temperature difference, and the heat load of thermosiphon evaporators

In the following Figures 10.9–10.13 it is shown how the driving height $H1$ influences:

- the circulating rate W through the thermosiphon evaporator,
- the pressure loss by thermosiphon circulation,
- the effective temperature difference LMTD and the heat load.

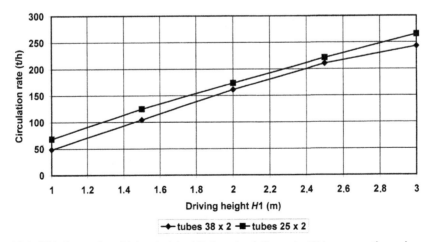

Figure 10.9 With increasing driving height $H1$ the circulating rate W increases through a vertical thermosiphon evaporator.

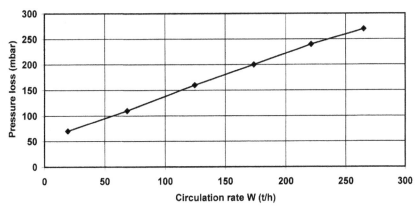

Figure 10.10 The pressure loss in the evaporator circulation increases with increasing circulating rate.

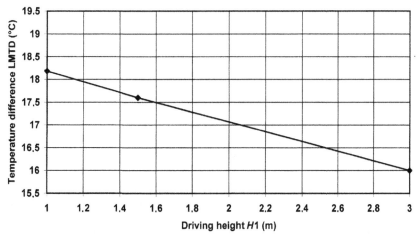

Figure 10.11 The effective temperature difference LMTD of a steam-heated vertical thermosiphon evaporator decreases with increasing driving height $H1$, because the higher hydrostatic pressure increases the boiling pressure.

10.6 DESIGN OF RISER AND DOWNCOMER DIAMETER

Recommended riser diameter D_R for a vertical thermosiphon evaporator:

$$D_R = d_i \times \sqrt{n} \, (m)$$

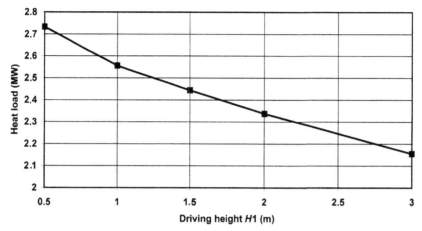

Figure 10.12 The heat load of a vertical thermosiphon evaporator decreases with increasing height $H1$ because the effective temperature difference LMTD decreases.

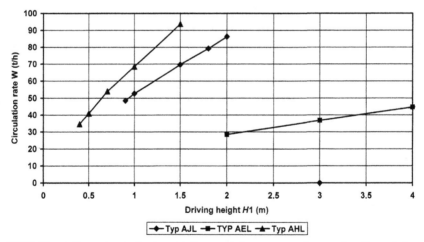

Figure 10.13 Circulating rates in horizontal thermosiphon evaporators as function of the driving height $H1$ for the evaporator types AJL, AHL, and AEL. Due to the high pressure loss in the evaporator type AEL a higher driving height $H1$ is needed.

Recommended minimum flow velocity of the two-phase flow in the riser to avoid pulsing piston flow with pressure variations in the column:

$$w_{min} = \sqrt{\frac{100}{\rho_V}}(m/s)$$

Recommended maximum flow velocity of the two-phase flow in the riser:

$$w_{max} = \sqrt{\frac{6000}{\rho_{2Ph}}}(m/s)$$

Recommended diameter D_D for the downcomer:

$$D_D = \frac{1}{2} \times D_R \ (m)$$

Recommended flow velocities: in the downcomer: 0.5−1 m/s:
d_i = evaporator tube inner diameter (m)
n = number of evaporator tubes
ρ_V = vapor density (kg/m³)
ρ_{2Ph} = density of the two-phase mixture (kg/m³)

Example 16: Determination of downcomer and riser diameter

$n = 211$ tubes $d_i = 21$ mm   Circulating rate $W = 40$ t/h
$\rho_{liq} = 750$ kg/m³ $\rho_{2Ph} = 24.3$ kg/m³ $\rho_V = 5$ kg/m³ $x = 0.2$
$V_{liq} = 53.3$ m³/h $V_{2Ph} = 1664$ m³/h

$$D_R = d_i \times \sqrt{n} = 0.021 \times \sqrt{211} = 0.305 \text{ m} \quad w_{2Ph} = 6.3 \text{ m/s}$$

$$D_D = \frac{1}{2} \times D_R = 0.5 \times 305 = 0.15 \text{ m} \qquad w_{Fl} = 0.84 \text{ m/s}$$

$$w_{min} = \sqrt{\frac{100}{\rho_V}} = \sqrt{\frac{100}{5}} = 4.5 \text{ m/s}$$

$$w_{max} = \sqrt{\frac{6000}{\rho_{2Ph}}} = \sqrt{\frac{6000}{24.3}} = 15.7 \text{ m/s}$$

Recommended nozzle spacing in the column
Height between the liquid head in the column and the two-phase inlet nozzle of the reboiler: ca. 0.5 m
    Height between the last bottom tray in the column and the two-phase inlet nozzle of the reboiler: 0.5−0.8 m
    Liquid height in the column bottom: 1−1.5 m

10.7 CALCULATION OF THE PRESSURE LOSSES IN THE THERMOSIPHON CIRCULATION

The total pressure loss in a thermosiphon circulation consists of the following (Figure 10.14):
- pressure loss in the downcomer for the liquid inflow
- friction pressure loss for the two-phase flow in the reboiler tubes
- acceleration pressure loss in the reboiler tubes
- friction pressure loss in the riser for the two-phase flow

10.7.1 Liquid pressure loss in the downcomer

$$\Delta P_{\text{Downc}} = f \times \frac{L}{d} \times \frac{w_{\text{Downc}}^2 \times \rho_{\text{liq}}}{2} \ (\text{Pa})$$

f = friction factor = $\frac{0.216}{\text{Re}^{0.2}}$

L = downcomer length (m)

d = diameter (m)

w_{Downc} = flow velocity in the downcomer (m/s)

ρ_{liq} = liquid density (kg/m^3)

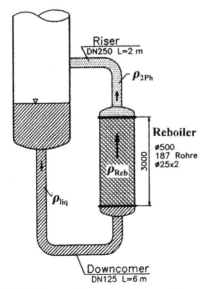

Figure 10.14 Pressure losses in a thermosiphon reboiler.

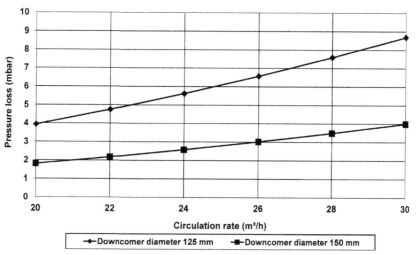

Figure 10.15 Pressure loss in the downcomer for two nominal widths as function of the circulating rate.

Example 17: Calculation of the pressure loss in the downcomer (Figure 10.15)

Downcomer length $L = 25$ m   Downcomer diameter $= 0.125$ m
Flow rate $= 30$ m³/h   Flow velocity $w_{Downc} = 0.68$ m/s
Kinematic viscosity $\nu = 0.4$ mm²/s   Reynolds number Re $= 212{,}314$
Friction factor $f = 0.0185$ Liquid density $\rho_F = 750$ kg/m³

$$\Delta P_{Downc} = 0.0185 \times \frac{25}{0.125} \times \frac{0.68^2 \times 750}{2} = 641 \text{ Pa} = 6.4 \text{ mbar}$$

10.7.2 Friction pressure loss for two-phase flow in the evaporator

According to Lockhart–Martinelli the fractional single-phase flow pressure loss of gas or liquid is converted to the pressure loss of the two-phase flow with the correction factors (Figure 10.16)

$$\Phi_V^2 \text{ or } \Phi_{liq}^2$$

$$\Delta P_{2ph} = \Phi_V^2 \times \Delta P_V = \Phi_{liq}^2 \times \Delta P_{liq}$$

$\Delta P_V =$ pressure loss for fractional vapor flow (Pa)
$\Delta P_{liq} =$ pressure loss for fractional liquid flow (Pa)

$$X = \sqrt{\frac{\Delta P_{liq}}{\Delta P_V}}$$

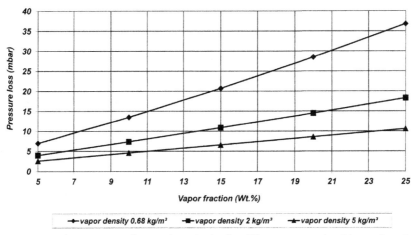

Figure 10.16 Two-phase pressure loss in the reboiler for different vapor densities as function of the vapor fraction in the two-phase mixture.

For the correction factors Φ_G^2 and Φ_F^2 it follows:

$$\Phi_{liq}^2 = 1 + \frac{C}{X} + \frac{1}{X^2} \quad \Phi_V^2 = 1 + C \times X + X^2$$

Turbulent liquid and gas flow: $C = 20$
Laminar liquid and turbulent gas flow: $C = 12$
Turbulent liquid and laminar gas flow: $C = 10$
Laminar liquid and gas flow: $C = 5$

10.7.3 Accelerated pressure loss in the evaporator

Due to the evaporation the flow volume increases and the flow in the evaporator tubes gets accelerated. The hereby caused pressure loss is calculated as follows:

$$\Delta P_{acc} = m^2 \times \left(\frac{1}{\rho_{2Ph}} - \frac{1}{\rho_{liq}} \right) = m \times (w_{out} - w_{in})(Pa)$$

$m = $ mass flow density $(kg/m^2\,s)$
$\rho_{2Ph} = $ density of the two-phase mixture
$\rho_{liq} = $ liquid density (kg/m^3)
$w_{in} = $ inlet flow velocity of the liquid (m/s)
$w_{out} = $ outlet flow velocity of the two-phase mixture (m/s)

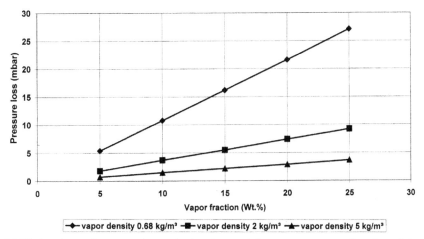

Figure 10.17 Accelerated pressure losses in the thermosiphon evaporator as function of the vaporization rate for different vapor densities.

Example 18: Calculation of the accelerated pressure loss

Incoming liquid rate: 22 t/h in 187 tubes 25 × 2 (Figure 10.17)

$$\rho_{liq} = 750 \text{ kg/m}^3 \qquad \rho_{2Ph} = 4.97 \text{ kg/m}^3 \qquad w_{in} = 0.125 \text{ m/s} \qquad w_{out} = 18.99 \text{ m/s}$$

$$m = \frac{22,000}{3600 \times 0.021^2 \times 0.785 \times 187} = 94.39 \text{ kg/m}^2 \text{ s}$$

$$\Delta P_{acc} = 94.39^2 \times \left(\frac{1}{4.97} - \frac{1}{750}\right) = 1781 \text{ Pa} = 17.8 \text{ mbar}$$

$$\Delta P_{acc} = 94.39 \times (18.99 - 0.125) = 1780 \text{ Pa} = 17.8 \text{ mbar}$$

10.7.4 Friction pressure loss in the riser for the two-phase flow

According to Lockhart–Martinelli the pressure loss of the fractional single-phase flow of gas or liquid is converted into the pressure loss of the two-phase flow with the correction factors Φ_V^2 or Φ_{liq}^2.

$$\Delta P_{2Ph} = \Phi_V^2 \times \Delta P_V = \Phi_{liq}^2 \times \Delta P_{liq}$$

$\Delta P_V =$ pressure loss for fractional vapor flow
$\Delta P_{liq} =$ pressure loss for fractional liquid flow

$$X = \sqrt{\frac{\Delta P_{liq}}{\Delta P_V}}$$

For the correction factors Φ_V^2 and Φ_{liq}^2 it follows:

$$\Phi_{liq}^2 = 1 + \frac{C}{X} + \frac{1}{X^2} \qquad \Phi_{liq}^2 = 1 + C \times X + X^2$$

Turbulent liquid and gas flow: $C = 20$
Laminar liquid and turbulent gas flow: $C = 12$
Turbulent liquid and laminar gas flow: $C = 10$
Laminar liquid and gas flow: $C = 5$

Example 19: Calculation of the two-phase pressure loss in the riser

Riser diameter: 250 mm   Equivalent length with elbow and reducers: $L = 30$ m

The two-phase mixture consists of 5 t/h vapor and 15 t/h liquid.

5 t/h vapor	$\rho_V = 2$ kg/m^3	$w_V = 14.15$ m/s	$\nu_V = 0.085$ mm^2/s
15 t/h liquid	$\rho_{liq} = 750$ kg/m^3	$w_{liq} = 0.113$ m/s	$\nu_{liq} = 0.85$ mm^2/s

Initially, the pressure losses for the liquid and the vapor flow are determined:

$$Re_{liq} = 33,235 \quad \text{Friction factor } f = \frac{0.216}{33,235^{0.2}} = 0.027$$

$$\Delta P_{liq} = 0.027 \times \frac{30}{0.25} \times \frac{0.113^2 \times 750}{2} = 15.4 \text{ Pa}$$

$$Re_V = 41,617 \quad \text{Friction factor } f = \frac{0.216}{41,617^{0.2}} = 0.0257$$

$$\Delta P_V = 0.0257 \times \frac{30}{0.25} \times \frac{14.15^2 \times 2}{2} = 617.5 \text{ Pa}$$

$$X = \sqrt{\frac{15.4}{617.5}} = 0.158$$

For turbulent gas and liquid flow $C = 20$.

$$\Phi_V^2 = 1 + 20 \times 0.158 + 0.158^2 = 4.18$$

$$\Phi_{liq} = 1 + \frac{20}{0.158} + \frac{1}{0.158^2} = 167.6$$

$$\Delta P_{2Ph} = 4.18 \times 617.5 = 167.6 \times 15.4 = 2581 \text{ Pa}$$

10.7.5 Total pressure loss in the thermosiphon circulation

The total pressure loss in the thermosiphon circuit is shown in Figure 10.18.

With increasing vapor rate the pressure loss in the thermosiphon circulation increases because due to the growing vapor fraction the flow velocity in the reboiler tubes increases.

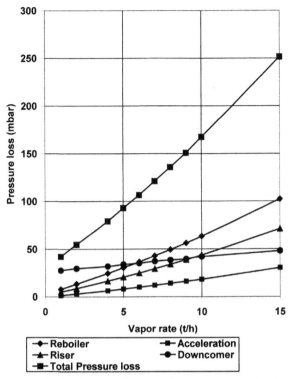

Figure 10.18 Pressure losses in the thermosiphon circulation at a vapor density of 5 kg/m³ as a function of the vapor rate at 20 t/h liquid circulation.

Figure 10.19 shows that the pressure loss at lower vapor densities is clearly larger than at higher vapor densities because with small vapor densities the volumetric fraction of the vapor is larger and hence also the flow velocity of the two-phase flow.

Design data:

Circulation rate: 20 t/h

Reboiler with 187 tubes 25 × 2, 3 m long

10.7.6 Reboiler characteristic curves

For the design of a thermosiphon circle it is recommended to create the so-called "reboiler characteristic curves." Therein the calculated pressure loss in the thermosiphon circulation is plotted against the liquid circulation. In the same diagram the available pressure potential for the liquid circulation at different driving heights $H1$ is plotted. From the reboiler characteristic curves it can be read off which elevation height $H1$ for a certain liquid circulation is required (Figure 10.20).

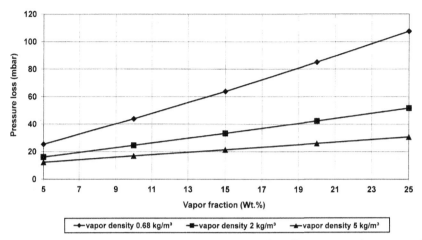

Figure 10.19 Pressure losses in the thermosiphon circulation for different vapor densities and 20 t/h circulation rate.

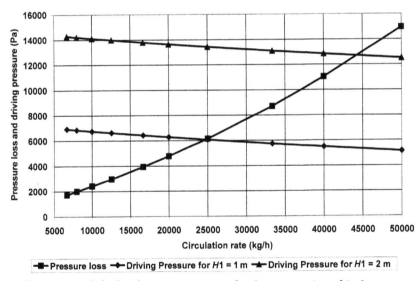

Figure 10.20 Reboiler characteristic curves for the evaporation of 2 t/h vapor.

10.8 CALCULATION OF THE REQUIRED REBOILER LENGTH OR AREA FOR THE HEATING UP TO THE BOILING TEMPERATURE AND FOR THE EVAPORATION IN VERTICAL THERMOSIPHON EVAPORATORS

In Figure 10.21 it is shown that in the thermosiphon evaporator, two processes occur: heating of the liquid up to the boiling temperature at the pressure in the evaporator and subsequently the actual evaporation. From point B to point C the liquid is heated

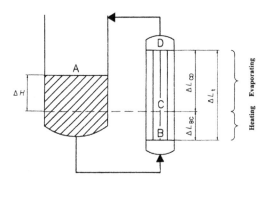

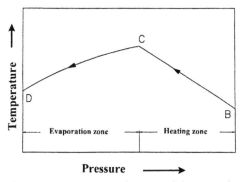

Figure 10.21 Heating and evaporation zones in the vertical thermosiphon reboiler with pressure and temperature curve.

up to the boiling temperature. The temperature rises and the pressure decreases because the static pressure is lower due to the superimposed liquid height at C. The evaporation starts at point C. From point C to point D the static pressure falls and the boiling temperature decreases.

10.8.1 Required reboiler length and area for heating up to the boiling temperature

$$L_{\text{heat}} = \frac{G_{\text{liq}} \times q_{\text{liq}} \times \Delta t_{\text{heat}}}{n \times d \times \pi \times U_{\text{heat}} \times \text{LMTD}_{\text{heat}}} \, (\text{m}) \quad A_{\text{heat}} = \frac{Q_{\text{heat}}}{U_{\text{heat}} \times \text{LMTD}_{\text{heat}}} \, (\text{m}^2)$$

G_{liq} = liquid inlet into the evaporator (kg/h)
q_{liq} = specific heat capacity of the liquid (Wh/kg K)
Δt_{heat} = required heating temperature up to the boiling point (°C)
n = number of tubes
d = tube diameter

U_{heat} = overall heat transfer coefficient for the convective heating (W/m² K)
LMTD$_{heat}$ = temperature difference for the heating (°C)
Q_{heat} = heat duty for the heating (W)

The required temperature difference for the heating up to the boiling temperature is initially estimated, for example, 3–10 °C.

The overall heat transfer coefficient U_{heat} for the heating is determined from the heat transfer coefficients α_{liq} for the heating of the liquid product and α_{steam} for condensing steam ($\alpha_{steam} \approx 5000$ W/m² K):

The heat transfer coefficient α_{liq} for the heating of the product in the tube is determined with the formula for the convective heat transfer in the tube.

Due to the low flow velocity at the inflow of the liquid into the evaporator ($w \approx 0.1$–0.25 m/s) the α_{liq}-values are very low:

$w = 0.1$ m/s	Re = 5250	Pr = 5.4	Nu = 32.81	$\alpha_{liq} = 187.5$ W/m² K
$w = 0.15$ m/s	Re = 7875	Pr = 5.4	Nu = 49.28	$\alpha_{liq} = 281$ W/m² K
$w = 0.2$ m/s	Re = 10,500	Pr = 5.4	Nu = 66.13	$\alpha_{liq} = 377.9$ W/m² K
$w = 0.25$ m/s	Re = 13,125	Pr = 5.4	Nu = 79.05	$\alpha_{liq} = 451.7$ W/m² K

Example 20: Required tube length for convective heating to the boiling point

$G_{liq} = 20,000$ kg/h $c_{liq} = 0.6$ Wh/kg K $n = 187$ tubes $d = 0.025$ m
Heating steam temperature $t_{steam} = 165$ °C   Boiling temperature = 155 °C
Product inlet temperature = 150 °C   Estimated value for $\alpha_{liq} = 350$ W/m² K
Required heating $\Delta t_{heat} = 155 - 150$ °C = 5 °C $\alpha_{steam} = 5000$ W/m² K

Product temperature	150	→	155
Steam temperature	165	≈	165
Differences	15		10

$$\text{LMTD}_{heat} = \frac{15 - 10}{\ln\frac{15}{10}} = 12.3\ °C \qquad \frac{1}{U_{heat}} = \frac{1}{350} + \frac{1}{5000} + 0.00025 \quad U_{heat} = 302\ W/m^2\ K$$

Chosen : $U_{heat} = 300$ W/m² K

$$L_{heat} = \frac{20,000 \times 0.6 \times (155 - 150)}{187 \times 0.025 \times \pi \times 300 \times 12.3} = 1.11\ m$$

$$A_{heat} = \frac{20,000 \times 0.6 \times 5}{300 \times 12.3} = 16.3\ m^2$$

Check: $A_{heat} = 1.11 \times 187 \times 0.025 \times \pi = 16.3\ m^2$

10.8.2 Required tube length L_{evap} and area A_{evap} for evaporation

$$L_{evap} = \frac{G_V \times r}{n \times d \times \pi \times U_{evap} \times LMTD_{evap}} \, (m)$$

$$A_{evap} = \frac{G_V \times r}{U_{evap} \times LMTD_{evap}} \, (m^2)$$

G_V = evaporated rate (kg/h) r = latent heat (Wh/kg)
U_{evap} = overall heat transfer coefficient for the evaporation (W/m² K)
$LMTD_{evap}$ = temperature difference for the evaporation (K)

Example 21: Calculation of the required tube length for the evaporation at 155 °C

G_V = 3000 kg/h r = 80 Wh/kg U_{evap} = 1000 W/m² K
Heating steam temperature: 165 ≈ 165 °C
Boiling temperature: 155 → 155 °C   LMTD = 10 °C

$$L_{evap} = \frac{G_V \times r}{n \times d \times \pi \times U_{evap} \times LMTD_{evap}} = \frac{3000 \times 80}{187 \times 0.025 \times \pi \times 1000 \times 10} = 1.63 \text{ m}$$

$$A_{evap} = \frac{G_V \times r}{U_{evap} \times LMTD_{evap}} = \frac{3000 \times 80}{1000 \times 10} = 24 \text{ m}^2$$

Cross-check: A_{evap} = 1.63 × 187 × 0.025 × π = 24 m²
Alternative calculation for U_{heat} = 180 W/m² K *and* U_{evap} = 800 W/m² K

L_{heat} = 1.84 m A_{heat} = 27.1 m²
L_{evap} = 2.04 m A_{evap} = 30 m²

The fractional area for the heating requires almost 50%!

10.9 REQUIRED HEATING LENGTH FOR VERTICAL THERMOSIPHON REBOILERS ACCORDING TO FAIR [2]

The required tube length fraction X_{heat} for the heating is calculated as follows (see Figure 10.22):

$$\frac{P_B - P}{P_B - P_A} = X = \frac{\Delta t / \Delta P}{\dfrac{\Delta t / \Delta L}{\Delta P / \Delta L} + \dfrac{\Delta t}{\Delta P}} \qquad \frac{\Delta P}{\Delta L} = \frac{\rho_{liq} \times g}{L} \, (mbar/m)$$

$$\frac{\Delta t}{\Delta L} = \frac{n \times d \times \pi \times LMTD_{heat} \times U_{heat}}{G_{liq} \times c_{liq}} \, (°C/m)$$

$\Delta t / \Delta L$ = heating gradient of the product per meter tube length L (°C/m)
$\Delta P / \Delta L$ = pressure increase due to the liquid height in the tube (mbar/m)
$\Delta t / \Delta P$ = temperature gradient of the vapor pressure curve (°C/mbar)

Table 10.1 Temperature gradients of the vapor pressure curve

Product	Temperature range (°C)	Temperature gradient (°C/mbar)
Benzene	50/70	0.054
	80/100	0.025
	94/104	0.021
Hexane	60/68	0.035
	70/80	0.027
	82/92	0.021
Toluene	90/100	0.05
	110/120	0.031
	126/136	0.022
Heptane	68/80	0.069
	90/100	0.036
	103/123	0.024
Octane	70/100	0.097
	120/130	0.036
	132/152	0.025

The value $\Delta t/\Delta P$ must be determined from the vapor pressure values.

In Table 10.1 the temperature gradients for some products are listed.

With increasing temperature or at higher pressures the temperature gradient decreases.

In vacuum the temperature gradient is larger and so a greater heating surface area for the heating up to the boiling temperature is required.

Example 22: Calculation of the fractional tube length X_{heat} for the heating

$n = 187$ tubes $d = 0.025$ m   LMTD $= 12.3$ °C $U_{heat} = 300$ W/m^2 K
$G_{liq} = 20{,}000$ kg/h $c_{liq} = 0.6$ Wh/kg K $\rho_{liq} = 750$ kg/m^3 $L_{tot} = 3$ m

$$\frac{\Delta t}{\Delta L} = \frac{187 \times 0.025 \times \pi \times 12.3 \times 300}{20{,}000 \times 0.6} = 4.51 \text{ °C/m}$$

$$\frac{\Delta P}{\Delta L} = \frac{750 \times 9.81}{1} = 7357 \text{ Pa/m} = 73.57 \text{ mbar/m}$$

$\Delta t/\Delta P = 0.03$ °C/mbar from the vapor pressure curve

$$X_{heat} = \frac{0.03}{\frac{4.51}{73.57} + 0.03} = 0.328$$

Required tube length for the heating $L_{heat} = X_{heat} \times L_{tot} = 0.328 \times 3 = 0.98$ m

The fractional length for the heating becomes shorter if the heating rate for $\Delta t/\Delta L$ (°C/m tube) is enlarged, for instance, by more tubes or a better overall heat transfer coefficient for the heating or more heating area by structured surfaces (fins) or by the installation of turbulators.

Example:
$\Delta t/\Delta L = 4.5$ °C/m ➜ $L_{heat} = 0.98$ m
$\Delta t/\Delta L = 12$ °C/m ➜ $L_{heat} = 0.47$ m

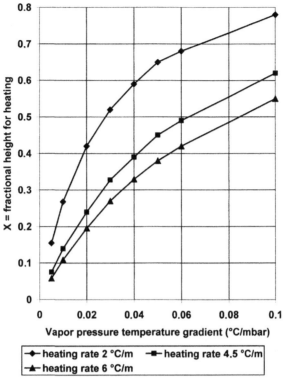

Figure 10.22 The fractional heating tube length as a function of the temperature gradient of the vapor pressure.

The *temperature gradient $\Delta t/\Delta P$ is especially large in the low-pressure region*. Therefore, in vacuum evaporators the required length for the heating is particularly large.

In Figure 10.22 the X_{heat}-value as the function of the heating gradient for different temperature gradients of the vapor pressure curve is shown. The X_{heat}-value increases with rising temperature gradient of the vapor pressure curve because more has to be heated.

The required heating area is reduced by a better heating.

Example 23: Determination of the required heating length for different heating gradients

Reboiler height $H4 = 3$ m

Temperature gradient of the vapor pressure curve $\Delta t/\Delta P = 2.2$ °C/m FS

$\Delta t/\Delta L = 2$ °C/m	$X_{heat} = 0.52$	$H_{heat} = 1.57$ m
$\Delta t/\Delta L = 4$ °C/m	$X_{heat} = 0.35$	$H_{heat} = 1.06$ m
$\Delta t/\Delta L = 6$ °C/m	$X_{heat} = 0.27$	$H_{heat} = 0.8$ m

10.10 CALCULATION OF THE PRESSURE AND BOILING POINT INCREASE BY MEANS OF THE LIQUID HEIGHT *H*1

The pressure in the evaporator is increased due to the required driving height *H*1 for the thermosiphon circulation and the required height for heating up to the boiling temperature.

In Figure 10.23 it is shown how the pressure in the reboiler rises with increasing driving height *H*1.

Due to the higher pressure above the evaporating area the boiling point becomes higher and the effective temperature difference for the heat transfer is reduced.

For the actual function of the reboiler, the evaporation, the heating wall temperature must lie above the boiling point. In multicomponent mixtures additionally the temperature rise from boiling point to dew point must be considered. Therefore the liquid circulation through the thermosiphon evaporator must be correspondingly heated taking the boiling point increase into account (Figure 10.24).

The calculations are performed as follows:

Initially the pressure is calculated at point C and the pressure increase between point A and C is determined. The required boiling point increase for this pressure rise is then determined.

P_A = pressure above the liquid in the column bottom (Pa) or (mbar)

P_B = pressure at the bottom in the evaporator

P_C = pressure above the liquid level in the reboiler

In Table 10.2 it is shown that a higher inflow height *H*1 increases the pressure in the evaporator and hence also the boiling point (Figure 10.23).

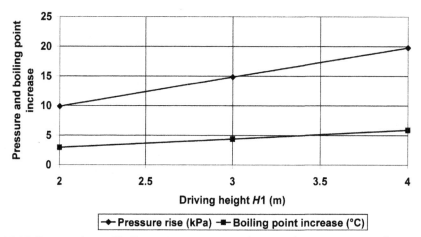

Figure 10.23 Pressure rise and boiling point increase in a thermosiphon evaporator as function of the driving height.

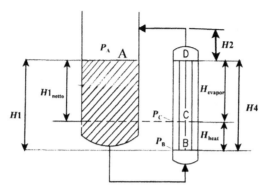

Figure 10.24 Pressures P_A, P_B, and P_C in the reboiler circuit.

Table 10.2 Boiling point increase at different H1-values

H1 (m)	H_{heat} (m)	P_B (mbar)	ΔP_{BC} (mbar)	P_C (mbar)	ΔP_{AC} (mbar)	Δt_{AC} (°C)
2	0.65	1147	47.8	1099	99	2.97
3	0.98	1220	72	1148	148	4.44
4	1.31	1294	96.3	1198	198	5.94

The required heating length is derived from the quotient boiling point increase divided by the temperature increase by heating per meter tube.

$$P_B = P_A + H1 \times \rho_{liq} \times 9.81 \,(\text{Pa}) \quad P_C = P_B - H_{heat} \times \rho_1 \times 9.81 \,(\text{Pa})$$
$$P_C = P_A + \Delta P + 9.81 \times [H2 \times \rho_{2Ph} + (H4 - H_{heat}) \times \rho_3] \,(\text{Pa})$$
$$\Delta P_{AC} = P_C - P_A \,(\text{Pa}) \quad \Delta P = \text{Pressure loss in the reboiler (Pa)}$$

Boiling point increase $\Delta t_{AC} = \Delta P_{AC} \times \Delta t / \Delta P$ (°C)
Required heating length $H_{heat} = \Delta t_{AC} / \Delta t / \Delta L$ (m)
In the normal case at atmospheric evaporation $H1 = H4$.
The liquid level in the column is at the level of the top tube sheet of the evaporator.
$H1_{netto} = H1 - H_{heat} = H1 \times (1 - X_{heat})$
$H4 = H_{heat} + H_{evap}$

Example 24: Calculation of the pressure and boiling point increase

$P_A = 1$ bar $H1 = 3$ m $H2 = 2$ m $H4 = 4$ m $H_{heat} = 1$ m
$\rho_{liq} = 750$ kg/m $\rho_{2Ph} = 47.1$ kg/m $\rho_{Reb} = 135.8$ kg/m $\Delta P = 9800$ Pa

$P_B = 100{,}000 + 3 \times 750 \times 9.81 = 122{,}072$ Pa
$P_C = 122{,}072 - 9.81 \times (1 \times 750) = 114{,}715$ Pa
$P_C = 100{,}000 + 9800 + 9.81 \times (2 \times 47.1 + 3 \times 135.8) = 114{,}720$ Pa
Pressure increase between A and C $= \Delta P_{AC} = 114{,}720 - 100{,}000 = 14{,}720$ Pa $= 147.2$ mbar
Temperature increase at pressure rise $\Delta t / \Delta P = 0.03$ °C/mbar
Boiling temperature increase between A and C $= \Delta t_{AC} = 0.03 \times 147.2 = 4.41$ °C

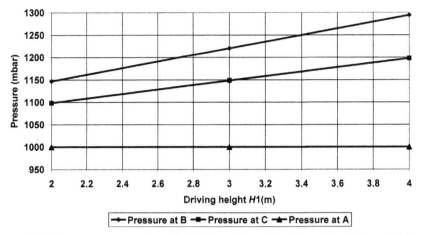

Figure 10.25 Pressure curve at the point A, B, and C as a function of the driving height H1.

Required heating length for $\Delta t/\Delta L = 4.5\,°C/m$ tube
$H_{heat} = 4.41/4.5 = 0.98$ m tube length

Basic data: $X_{Heiz} = 0.328$ $\Delta t/\Delta P = 0.03\,°C/mbar$ $\Delta t/\Delta L = 4.5\,°C/m$ tube

The pressures at the points A, B, and C as a function of the driving height H1 is shown in Figure 10.25.

In *evaporations in vacuum* the driving height H1 is reduced to 50% of the reboiler height H4 in order to reduce the pressure increase in the evaporator and the required heating height. The liquid level in the column is at half of the reboiler height. The following Example 25 makes it clear.

Example 25: Boiling point increase and required heat length at vacuum evaporation

Heating gradient $\Delta t/\Delta L = 0.054\,°C/mbar = 4\,°C/m$ tube length
Temperature gradient of the vapor pressure curve $\Delta t/\Delta P = 0.06796\,°C/mbar = 5\,°C$ pro m FS

$$X_{heat} = \frac{5}{4+5} = 0.555$$

Case 1:   H1 = 3 m   H4 = 3 m $H_{heat} = X_{heat} \times H4 =$
                                                                     $0.555 \times 3 = 1.67$ m

$\rho_{liq} = 750\,kg/m^3$ $\rho_{2Ph} = 25.07\,kg/m^3$ $P_A = 100\,mbar = 10,000\,Pa$

$P_B = 10,000 + 9.81 \times 3 \times 750 = 32,072\,Pa$
$P_C = 32,072 - 9.81 \times 1.67 \times 750 = 19,785\,Pa$
$\Delta P_{AC} = 19,785 - 10,000 = 9785\,Pa = \frac{9785}{9.81 \times 750} = 1.33$ m FS = 97.85 mbar
Temperature increase due to the pressure rise of 97.85 mbar:

$\Delta t_{AC} = 0.06796 \times 97.85 = 6.65\,°C$ $\Delta t_{AC} = 5 \times 1.33 = 6.65\,°C$

$$H_{heat} = \frac{\Delta t_{AC}}{\Delta t/\Delta L} = \frac{6.65(°C)}{4(°C/m \text{ tube length})} = 1.66\,m$$

Case 2: $H1 = 1.5$ m $H_{heat} = X_{hat} \times H1 = 0.555 \times 1.5 = 0.83$ m

$P_B = 10,000 + 9.81 \times 1.5 \times 750 = 21,036$ Pa
$P_C = 21,036 - 9.81 \times 0.83 \times 750 = 14,930$ Pa
$\Delta P_{AC} = 14,930 - 10,000 = 4930$ Pa $= 49.3$ mbar $= 0.67$ m FS
Temperature increase by the pressure rise of 49.3 mbar:

$$\Delta t_{AC} = 0.06796 \times 49.3 = 3.35 \,°C \qquad \Delta t_{AC} = 5 \times 0.67 = 3.35 \,°C$$

$$H_{heat} = \frac{\Delta t_{AC}}{\Delta t / \Delta L} = \frac{3.35(°C)}{4(°C/m \text{ tube length})} = 0.83 \text{ m}$$

10.11 AVERAGE OVERALL HEAT TRANSFER COEFFICIENT FOR HEATING + VAPORIZING

For the calculation of an average overall heat transfer coefficient U_a the following is valid with X_{heat}:

$$U_a = X_{heat} \times U_{heat} + (1 - X_{heat}) \times U_{evap} \left(W/m^2 \, K \right)$$

Since the driving temperature differences for the heating and the vaporizing are different, the calculation of an average overall heat transfer coefficient is not quite correct.

Example 26: Calculation of the average overall heat transfer coefficient U_a

$$X_{heat} = 0.4 \qquad U_{heat} = 300 \text{ W/m}^2 \text{ K} \qquad U_{evap} = 2000 \text{ W/m}^2 \text{ K}$$

$U_a = X_{heat} \times U_{heat} + (1 - X_{heat}) \times U_{evap} = 0.4 \times 300 + 0.6 \times 2000 = 1320$ W/m^2 K
The good U-value for the vaporizing is substantially deteriorated by the heating.

10.12 CALCULATION OF THE VAPOR FRACTION X OF THE TWO-PHASE MIXTURE IN A VERTICAL REBOILER

The evaporated product rate changes above the height of the evaporator. At the inlet of the liquid in the evaporator the vapor fraction $x = 0$. The calculation of the vapor fraction in different heights L is carried out as follows:

$$x = \frac{q \times L \times n \times d \times \pi}{r \times W}$$

$$G_V = x \times W \, (\text{kg/h})$$

$$G_V = \frac{q \times A}{r} = \frac{q \times (n \times d \times \pi \times L)}{r} \, (\text{kg/h})$$

W = circulating rate through the thermosiphon evaporator (kg/h)

G_V = vapor rate (kg/h)

q = heat flux density (W/m²) = $U \times \Delta t$

r = heat of vaporization (Wh/kg)

x = vapor content in the two-phase mixture (fraction)

n = number of tubes

d = tube diameter (m)

L = tube length for the evaporation (m)

The x-values at different heights are required for the determination of the heat transfer coefficients and the pressure losses at the different zones of the evaporator. The calculation should be subdivided into several zones because the conditions change strongly above the height due to the different vapor content x in the two-phase mixture.

Example 27: Calculation of the vapor content at different tube lengths

$$q = 10{,}000 \text{ W/m}^2 \qquad n = 187 \text{ tubes} \qquad d = 0.025 \text{ m} \qquad r = 80 \text{ Wh/kg}$$
$$L = 1.63 \text{ m} \qquad W = 20{,}000 \text{ kg/h}$$

$$x = \frac{q \times L \times n \times d \times \pi}{r \times W} = \frac{10{,}000 \times 1.63 \times 187 \times 0.025 \times \pi}{80 \times 20{,}000} = 0.15$$

$$G_V = x \times W = 0.15 \times 20{,}000 = 3000 \ (\text{kg/h})$$

$$G_V = \frac{q \times A}{r} = \frac{q \times (n \times d \times \pi \times L)}{r} (\text{kg/h}) = \frac{10{,}000 \times 0.025 \times 187 \times 1.63 \times \pi}{80} = 3000 \ (\text{kg/h})$$

With these equations the vapor content x in the different heights of the evaporator above the heating length can be determined:

$L = 0.5$ m	$x = 0.0458$	916 kg/h vapor
$L = 1.0$ m	$x = 0.0918$	1836 kg/h vapor
$L = 1.4$ m	$x = 0.128$	2560 kg/h vapor
$L = 1.63$ m	$x = 0.150$	3000 kg/h vapor

In Figure 10.26 it is shown how the two-phase density and the flow velocity of the two-phase mixture change with increasing vaporization.

10.13 THERMOSIPHON REBOILER DESIGN EXAMPLE

Heat load $Q = 600$ kW $r = 150$ Wh/kg $c = 0.6$ Wh/kg °C

Required heating $\Delta t = 4$ °C $x = 0.2$ kg/kg vapor fraction

Vapor density $\rho_D = 5$ kg/m³   Liquid density $\rho_F = 750$ kg/m³

Calculation of the circulating rate for $x = 0.2$:

$$W = \frac{600{,}000}{0.6 \times 4 + 0.2 \times 150} = 18{,}518 \text{ kg/h}$$

$$G_V = 0.2 \times 18{,}518 = 3704 \text{ kg/h}$$

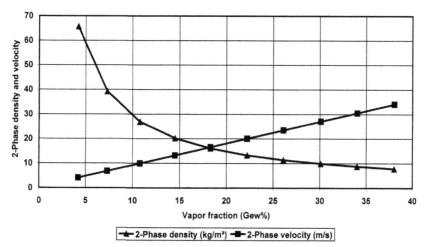

Figure 10.26 Two-phase density and flow velocity as a function of the evaporation rate.

Thermosiphon circulation $W = 18,518$ kg/h
Evaporation rate $D = 3704$ kg/h vapors

Circulating rates at other evaporation rates x:

$x = 0.1$	$W = 34,483$ kg/h	$D = 3448$ kg/h
$x = 0.15$	$W = 24,096$ kg/h	$D = 3614$ kg/h
$x = 0.2$	$W = 18,518$ kg/h	$D = 3704$ kg/h
$x = 0.25$	$W = 15,038$ kg/h	$D = 3759$ kg/h
$x = 0.3$	$W = 12,658$ kg/h	$D = 3797$ kg/h

Estimation of the *reboiler area A*:

$$U_{est} = 700 \text{ W/m}^2 \text{ K} \qquad \Delta t = 15 \,°C \qquad L = 3.5 \text{ m} \qquad \text{Tubes } 25 \times 2$$

$$A = \frac{600,000}{700 \times 15} = 57.1 \text{ m}^2$$

$$n = \frac{57.1}{\pi \times 0.025 \times 3.5} = 207 \text{ tubes } 25 \times 2$$

Chosen: 211 tubes with triangular pitch 32 mm
Shell inner diameter $D_i = 483.2$ mm for 211 tubes with pitch 32 mm
Reboiler area $A = n \times d_a \times \pi \times L = 211 \times 0.025 \times \pi \times 3.5 = 58 \text{ m}^2$

Calculation of the *two-phase density and the reboiler density* and the flow velocities in the reboiler:

$$\rho_{2Ph} = \frac{1}{\frac{x}{\rho_G} + \frac{1-x}{\rho_F}} = \frac{1}{\frac{0.1}{5} + \frac{0.9}{750}} = 47.1 \text{ kg/m}^3$$

$$\rho_3 = \frac{1}{\frac{1}{47.1} - \frac{1}{750}} \times \ln\left[\frac{\frac{1}{47.1} - \frac{1}{750}}{\frac{1}{750}}\right] = 135.8 \text{ kg/m}^3$$

Two-phase and reboiler densities for other x-values:

x (kg/kg)	ρ_{2Ph} (kg/m^3)	ρ_3 (kg/m^3)
0.1	47.1	135.8
0.15	32.1	104.2
0.2	24.3	85.3
0.25	19.6	72.8
0.3	16.4	63.7

Flow velocity in the 211 tubes for $W = 34{,}483$ kg/h and $x = 0.1$:

Flow cross-section $a_t = 211 \times 0.021^2 \times \pi/4 = 0.07308 \text{ m}^2$

Liquid velocity at the inlet:

$$w_{liq} = \frac{34{,}483}{750 \times 0.07308 \times 3600} = 0.175 \text{ m/s}$$

Two-phase flow velocity at the outlet:

$$w_{2Ph} = \frac{34{,}483}{47.1 \times 0.07308 \times 3600} = 2.78 \text{ m/s}$$

Average flow velocity in the reboiler:

$$w_m = \frac{34{,}483}{135.8 \times 0.07308 \times 3600} = 0.96 \text{ m/s}$$

Calculation of the heat transfer and overall heat transfer heat coefficient

Heat transfer coefficient for *convective heat transfer*

$$w_{liq} = 0.175 \text{ m/s} \qquad Re = 9187 \qquad Pr = 5.4$$

$$Nu = 0.023 \times Re^{0.8} \times Pr^{0.33} = 0.023 \times 9187^{0.8} \times 5.4^{0.33} = 59.42$$

$$\alpha_{liq} = \frac{Nu \times \lambda}{d_i} = \frac{59.42 \times 0.12}{0.021} = 339.57$$

$$X_{tt} = \left(\frac{1 - 0.1}{0.1}\right)^{0.9} \times \left(\frac{5}{750}\right)^{0.5} \times \left(\frac{0.3}{0.012}\right)^{0.1} = 0.8139$$

$$M = 2.5 \times \left(\frac{1}{0.8139}\right)^{0.7} = 2.88$$

$$\alpha_{conv} = 2.88 \times 339.6 = 980.6 \left(\text{W/m}^2 \text{ K}\right)$$

Heat *transfer coefficient for the boiling* of mixtures according to Mostinski
$\alpha_{NB} = 822 \text{ W/m}^2 \text{ K}$

$$Re_{2Ph} = 9187 \times 2.88^{1.25} = 34,468$$

$$S = \frac{1}{1 + (2.52 \times 10^{-6} \times 34,468^{1.17})} = 0.66$$

$$\alpha_n = 0.66 \times 822 = 542.7 \text{ W/m}^2 \text{ K}$$

Total heat transfer coefficient for the evaporation:
$\alpha_{Reb} = 980.6 + 542.7 = 1523 \text{ W/m}^2 \text{ K}$
Conversion of α_i to the outer tube surface area:

$$\alpha_{io} = \alpha_i \times \frac{d_i}{d_a} = 1523 \times \frac{21}{25} = 1279 \text{ W/m}^2 \text{ K}$$

Calculation of the overall heat transfer coefficient U for steam heating with
$\alpha_{steam} = 7000 \text{ W/m}^2 \text{ K}$:

$\alpha_{io} = \alpha_{Reb} = 1279 \text{ W/m}^2 \text{ K} \quad \alpha_{steam} = 7000 \text{ W/m}^2 \text{ K} \quad \Sigma f = 0.0002$

$$\frac{1}{U} = \frac{1}{\alpha_{Reb}} + \frac{1}{\alpha_{steam}} + \Sigma f = \frac{1}{1279} + \frac{1}{7000} + 0.0002 \quad U = 889 \text{ W/m}^2 \text{ K}$$

The determined *overall heat transfer coefficient* $U = 889 \text{ W/m}^2 \text{ K}$ for the evaporation is
larger then the estimated U-value $U_{est} = 700 \text{ W/m}^2 \text{ K}$ but it must also be heated up.
Cross-checking of the heat duty Q:

$Q = U \times A \times \Delta t = 889 \times 58 \times 15 = 773,452 \text{ W}$ $\qquad$ Reserve = 29%
Heating duty $Q_{heat} = 34,483 \times 0.6 \times 4 = 82,759 \text{ W}$ $\qquad$ LMTD = 19 °C

Required *heating area* A_{Heiz} for $U_{heat} = 340 \text{ W/m}^2 \text{ K}$ $\qquad$ $A_{Heiz} = \frac{82,759}{304 \times 19} = 14.3 \text{ m}^2$
Vaporization duty $Q_{verd} = 3448 \times 150 = 517,200 \text{ W}$ $\qquad$ LMTD = 15 °C
Required *vaporization area* A_{evap} for $U_{Reb} = 889 \text{ W/m}^2 \text{ K}$ $\qquad$ $A_{Verd} = \frac{517,200}{889 \times 15} = 38.8 \text{ m}^2$
Total required area $A_{req} = 14.3 + 38.8 = 53.1 \text{ m}^2 < 58 \text{ m}^2$ $\quad$ 9% reserve!

Recommendation: Enlargement of the area by longer tubes in order to improve the
reserve.

$L = 4 \text{ m length} \qquad A = 66.3 \text{ m}^2 \qquad 25\% \text{ reserve!}$

Calculation of the *required driving height H1* for an assumed pressure loss $\Delta P = 40$ mbar
in the thermosiphon circulation by downcomer, evaporator, and riser:

Riser height $H2 = 2 \text{ m}$ $\qquad$ Reboiler height $H4 = 4 \text{ m}$
$\rho_{liq} = 750 \text{ kg/m}^3$ $\qquad$ $\rho_{2Ph} = 47.1 \text{ kg/m}^3$ $\qquad$ $\rho_{Reb} = 135.8 \text{ kg/m}^3$

$$H1 = \frac{2 \times 4000 + 9.81 \times (2 \times 47.1 + 4 \times 135.8)}{9.81 \times 750} = 1.94 \text{ m}$$

Chosen: $H1 = 2$ m

Allowable pressure loss ΔP_{zul}

$\Delta P_{allow} = 9.81 \times (2 \times 750 - 2 \times 47.1 - 4 \times 135.8) = 8462$ Pa $= 84.6$ mbar

Under *consideration of the heating height* H_{Heiz} in the reboiler:

If a part of the tube length in the evaporator is required for the heating, the heating length H_{heiz} must be considered when determining the h eight $H1$.

$$H4 = 4 \text{ m} \qquad H2 = 2 \text{ m} \qquad H_{heat} = 0.5 \text{ m}$$

$$H1 = \frac{S \times \Delta P + g \times (H2 \times \rho_{2Ph} + \rho_{Reb} \times (H4 - H_{heat}) + H_{heat} \times \rho_1}{g \times \rho_1}$$

$$H1 = \frac{2 \times 4000 + 9.81 \times (47.1 \times 2 + 135.8 \times 3.5 + 0.5 \times 750)}{9.81 \times 750} = 2.35 \; m$$

The required height $H1$ becomes higher.

Cross-checking the allowable pressure loss:

$\Delta P_{allow} = g \times (H1 \times \rho_{liq} - H2 \times \rho_{2ph} - (H4 - H_{heat}) \times \rho_{Reb} - H_{heat} \times \rho_{liq})$

$\Delta P_{zul} = 9.81 \times (2.35 \times 750 - 2 \times 47.1 - 3.5 \times 135.8 - 0.5 \times 750) =$
8025 Pa $= 80.2$ mbar

Calculation of the *diameters for downcomer and riser for* $W = 34{,}483$ kg/h:

$$D_{Riser} = 0.21 \times \sqrt{211} = 0.3 \text{ m} \quad \text{Chosen}: \; D = 0.25 \text{ m}$$

$$D_{Downc} = 0.5 \times 0.3 = 0.15 \text{ m}$$

$$w_{Riser} = \frac{732}{3600 \times 0.25^2 \times 0.785} = 4.14 \text{ m/s}$$

$$w_{Downc} = \frac{46}{3600 \times 0.15^2 \times 0.785} = 0.72 \text{ m/s}$$

$$w_{min \; Riser} = \sqrt{\frac{100}{5}} = 4.47 \text{ m/s} \quad w_{max \; Riser} = \sqrt{\frac{6000}{47.1}} = 11.3 \text{ m/s}$$

REFERENCES

[1] N.P. Liebermann, E.T. Liebermann, Working Guide to Process Equipment, McGraw-Hill, N.Y., 2003.
[2] J.R. Fair, What you need to design thermosiphon reboilers, Petr. Ref. 39 (2) (1960).
[3] G.A. Hughmark, Designing Thermosiphon Reboilers, CEP 57, 1961, 7, CEP 60 (1964), 7 und CEP 65 (1969), 7.
[4] J.C. Chen, Correlation for boiling heat transfer to saturated liquids in convective flow, Ind. Eng. Proc. Des. Dev. 58 (3) (1966).
[5] D.L. Love, No hassle reboiler selection, Hydroc. Process. (October 1992).
[6] E. Chen, Optimize reboiler design, Hydroc. Process. (July 2001).

[7]   G.R. Martin, A.W. Sloley, Effectively design and simulate thermosyphon reboiler systems, Hydroc. Process. (June–July 1995).
[8]   J.R. Fair, A. Klip, Thermal design of horizontal reboilers, Chem. Eng. Progr. 79 (3) (1983).
[9]   G.K. Collins, Horizontal thermosiphon reboiler design, Chem. Eng. (July 1976).
[10] R. Kern, How to design piping for reboiler systems, Chem. Eng. (June 1975).

CHAPTER 11

Double Pipe, Helical Coil, and Cross Flow Heat Exchanger

Contents

11.1 DOUBLE PIPE AND MULTIPIPE HEAT EXCHANGERS

The calculations of the heat transfer coefficient follow the equations for the heat transfer in tubes.

On the shell side, the different hydraulic diameters for the determination of the heat transfer coefficient d_h and the pressure loss d_h' must be considered.

Double pipe heat exchangers are applied for low heat loads, small volume flows, and high pressures (Figure 11.1).

Multitube heat exchangers with a pipe shell and several inner tubes without baffles have a more exchange area than the double pipe equipments.

11.1.1 Calculation of double pipe and multipipe heat exchangers [1,2]

The calculations are according to the equations for the heat transfer in the tube.

Laminar: $\mathrm{Re} < 2300$

$$\mathrm{Nu} = 1.86 \times \mathrm{Re}^{0.33} \times \mathrm{Pr}^{0.33} \times \left(\frac{d_i}{L}\right)^{0.33}$$

Intermediate region: $2300 < \mathrm{Re} < 8000$

$$\mathrm{Nu} = \left(0.037 \times \mathrm{Re}^{0.75} - 6.66\right) \times \mathrm{Pr}^{0.42}$$

© 2016 Elsevier Inc.
All rights reserved.

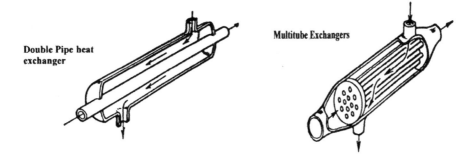

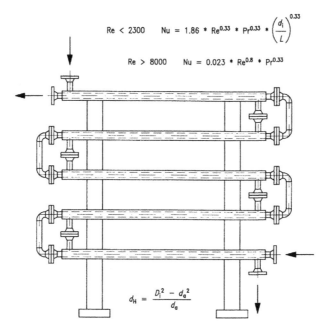

Figure 11.1 Double pipe heat exchanger.

Turbulent: $\mathrm{Re} > 8000$

$$\mathrm{Nu} = 0.023 \times \mathrm{Re}^{0.8} \times \mathrm{Pr}^{0.33} \qquad \mathrm{Pr} = \frac{\nu \times c \times \rho \times 3600}{\lambda}$$

$$\alpha_i = \frac{\mathrm{Nu} \times \lambda}{d_i} \left(\mathrm{W/m^2\,K}\right)$$

Flow velocity and Reynolds number in the tube:

$$w_t = \frac{V_t}{n \times d_i^2 \times \frac{\pi}{4}} \ (m/s) \qquad Re = \frac{w_t \times d_i}{\nu}$$

Flow cross-section a_{shell} and flow velocity w_{shell} in the annulus or shell:

$$a_{shell} = \frac{\pi}{4} \times \left(D_i^2 - n \times d_o^2\right) (m^2) \qquad w_{shell} = \frac{V_{shell}}{a_{shell} \times 3600} \ (m/s)$$

For the calculation of the heat transfer coefficient and the pressure loss, the *hydraulic diameter d_h for heat exchange and d_h' for pressure loss are needed.*
Reynolds number for the heat transfer with d_h: $Re = \dfrac{w_{shell} \times d_h}{\nu}$
Hydraulic diameter for double tube heat exchanger:

$$d_h = \frac{D_i^2 - n \times d_o^2}{n \times d_o} \ (m)$$

Hydraulic diameter for multitube heat exchanger:

$$d_h = \frac{D_i^2 - d_o^2}{d_o} \ (m)$$

Reynolds number Re' for the pressure loss with d_h': $Re' = \dfrac{w_{shell} \times d_h'}{\nu}$
Double tube heat exchanger: $d_h' = D_i - d_o$

Multitube heat exchanger: $d_h' = \dfrac{D_i^2 - n \times d_o^2}{D_i + n \times d_o}$
Pressure loss calculation:

$$\text{Nozzle pressure loss} \quad \Delta P_{nozz} = 1.5 \times \frac{w_{nozz} \times \rho}{2}$$

$$\text{Tube-side pressure loss} \quad \Delta P_t = \left(f \times \frac{L}{d_i} + z\right) \times \frac{w_t^2 \times \rho}{2} \ (Pa)$$

$$\text{Shell-side pressure loss} \quad \Delta P_{shell} = f \times \frac{L}{d_h'} \times \frac{w_{shell}^2 \times \rho}{2} \ (Pa)$$

$$\text{Friction factor} \quad f = \frac{0.275}{Re^{0.2}}$$

11.1.2 Calculation of the overall heat transfer coefficient U and driving temperature gradient LMTD and the required heat exchanger area A

$$\frac{1}{U} = \frac{1}{\alpha_{io}} + \frac{1}{\alpha_o} + \frac{s}{\lambda_{wall}} + f_i + f_o$$

$$\text{LMTD} = \frac{\Delta t_1 - \Delta t_2}{\ln \dfrac{\Delta t_1}{\Delta t_2}} (°C) \qquad A = \frac{Q}{U \times \text{LMTD}} \ (m^2)$$

Example 1: Double pipe heat exchanger design

Heat load $Q = 3.53$ kW
Inner tube 21.3×2 ($d_i = 17.3$ mm)
$V_T = 0.05$ m^3/h
$\rho = 1000$ kg/m^3
$\nu = 0.41$ mm^2/s
$c = 1.16$ Wh/kg K
$\lambda = 0.66$ W/m K
Pr $= 2.594$

LMTD $= 32.46$ °C
Shell pipe 33.7×2 ($D_i = 29.7$ mm)
$V_{shell} = 0.3$ m^3/h
$\rho = 1000$ kg/m^3
$\nu = 0.8$ mm^2/s
$c = 1.16$ Wh/kg K
$\lambda = 0.61$
Pr $= 5.476$

Tube side:

$$w_T = \frac{0.05}{3600 \times 0.0173^2 \times \pi/4} = 0.059 \text{ m/s}$$

$$\text{Re} = \frac{0.059 \times 0.0173}{0.41 \times 10^{-6}} = 2490$$

$$\text{Nu} = \left(0.037 \times 2490^{0.75} - 6.66\right) \times 2.594^{0.42} = 9.52$$

$$\alpha_i = \frac{9.52 \times 0.66}{0.0173} = 363 \text{ W/m}^2 \text{ K} \qquad \alpha_{io} = 363 \times \frac{17.3}{21.3} = 295 \text{ W/m}^2 \text{ K}$$

Shell side:

$$a_{shell} = \frac{\pi}{4} \times \left(0.0297^2 - 1 \times 0.0213^2\right) = 336.3 \times 10^{-6} \text{ m}^2$$

$$d_h = \frac{0.0297^2 - 0.0213^2}{0.0213} = 0.0201 \qquad \text{Re} = \frac{0.248 \times 0.0201}{0.8 \times 10^{-6}} = 6230$$

$$\text{Nu} = \left(0.037 \times \text{Re}^{0.75} - 6.66\right) \times 5.476^{0.42} = 39.39$$

$$\alpha_o = \frac{39.39 \times 0.61}{0.0201} = 1195 \text{ W/m}^2\text{ K}$$

Overall heat transfer coefficient U:

$$\frac{1}{U} = \frac{1}{1195} + \frac{1}{295} + \frac{0.002}{14} + 0.0002 = 0.00457 \qquad U = 219 \text{ W/m}^2\text{ K}$$

$$A_{req} = \frac{3530}{32.46 \times 219} = 0.5 \text{ m}^2 \qquad L_{req} = \frac{0.5}{\pi \times 0.0213} = 7.5 \text{ m pipe}$$

Pressure loss calculation for 2 × 3.75 m long tubes:
Tube side:

$$\Delta P_{nozz} = 1.5 \times \frac{0.059^2 \times 1000}{2} = 2.6 \text{ Pa} \qquad f = \frac{0.275}{2490^{0.2}} = 0.058$$

$$\Delta P_t = \left(0.058 \times \frac{7.5}{0.0173} + 1\right) \times \frac{0.059^2 \times 1000}{2} = 46 \text{ Pa}$$

$$\Delta P_{tot} = 2.6 + 46 = 48.6 \text{ Pa}$$

Shell side:

2 inlet and 2 outlet nozzles DN 20 $\qquad w_{nozz} = 0.265$ m/s for 0.3 m^3/h

$$\Delta P_{nozz} = 3 \times \frac{0.265^2 \times 1000}{2} = 105 \text{ Pa}$$

$$d'_h = D_i - d_a = 0.0297 - 0.0213 = 0.0084 \text{ m}$$

$$Re' = \frac{0.248 \times 0.0084}{0.8 \times 10^{-6}} = 2604$$

Friction factor $f = \dfrac{0.275}{Re^{0.2}} = \dfrac{0.275}{2604^{0.2}} = 0.057$

$$\Delta P_{shell} = \left(0.057 \times \frac{7.5}{0.0084} + 1\right) \times \frac{0.248^2 \times 1000}{2} = 1596 \text{ Pa}$$

$$\Delta P_{tot} = 105 + 1596 = 1701 \text{ Pa}$$

Example 2: Shell-side heat transfer coefficient in a multitube heat exchanger

$$V_{shell} = 2.3 \text{ m}^3/\text{h} \qquad Pr = 7.1 \qquad \lambda = 0.58 \text{ W/m K} \qquad \nu = 1 \text{ mm}^2/\text{s}$$

Shell diameter $D_i = 84$ mm with seven inner pipes 20×2, 3 m long

$$a_{shell} = \frac{\pi}{4} \times \left(0.084^2 - 7 \times 0.02^2\right) = 0.0033 \text{ m}^2$$

$$d_h = \frac{0.084^2 - 7 \times 0.02^2}{7 \times 0.02} = 0.0304 \text{ m} \quad d'_h = \frac{0.084^2 - 7 \times 0.02^2}{0.084 + 7 \times 0.02} = 0.019 \text{ m}$$

$$w_{shell} = \frac{2.3}{0.0033 \times 3600} = 0.194 \text{ m/s} \quad Re = \frac{0.194 \times 0.0304}{1 \times 10^{-6}} = 5897$$

Calculation of the heat transfer coefficient on the shell side with $Re = 5897$:

$$Nu = \left(0.037 \times 5897^{0.75} - 6.66\right) \times 7.1^{0.42} = 41.5$$

$$\alpha_o = \frac{Nu \times \lambda}{d_h} = \frac{41.5 \times 0.58}{0.0304} = 792 \text{ W/m}^2 \text{ K}$$

Example 3: Comparison of double pipe or multitube heat exchangers

Calculation of the double pipe heat exchanger.

Tube side:

5.2 m³/h	20/30 °C	$Q = 51,480$ W
$\rho = 1100$ kg/m³	$\lambda = 0.4$ W/m K	$c = 0.9$ Wh/kg K
$\nu = 2.4$ mm²/s	$w_T = 1.248$ m/s	$Pr = 21.38$
Inner pipe 42.4×2 mm	$d_i = 38.4$ mm	

$$Re = \frac{1.248 \times 0.0384}{2.4 \times 10^{-6}} = 19,968 \quad Nu = 0.023 \times 19,968^{0.8} \times 21.38^{0.33} = 174.14$$

$$\alpha_i = \frac{174.14 \times 0.4}{0.0384} = 1814 \text{ W/m}^2\text{K} \quad \alpha_{io} = 1814 \times \frac{38.4}{42.4} = 1643 \text{ W/m}^2 \text{ K}$$

Shell side:

$V_{shell} = 5.2$ m³/h	80/71.25 °C	$Pr = 2.35$
$\lambda = 0.666$ W/m K	$\nu = 0.385$ mm²/s	$c = 1.16$ Wh/kg K
Outer tube 70×2	$d_i = 66$ mm	$\rho = 974.65$ kg/m³

$$a_{shell} = \frac{\pi}{4} \times \left(0.066^2 - 0.0424^2\right) = 0.002 \text{ m}^2 \quad w_{shell} = 0.719 \text{ m/s}$$

$$d_h = \frac{0.066^2 - 0.0424^2}{0.0424} = 0.06033 \text{ m} \quad Re = \frac{0.719 \times 0.06033}{0.385 \times 10^{-6}} = 112,668$$

$$Nu = 335.4 \quad \alpha = \frac{335.4 \times 0.666}{0.06033} = 3703 \text{ W/m}^2 \text{ K}$$

Calculation of the overall heat transfer coefficient U:

$$\frac{1}{U} = \frac{1}{1643} + \frac{1}{3703} + \frac{0.002}{50} + 0.0002 = 0.001119 \qquad U = 894 \text{ W/m}^2 \text{ K}$$

$$\text{LMTD} = 50.6 \,°\text{C}$$

$$A_{\text{req}} = \frac{51,480}{894 \times 50.6} = 1.138 \text{ m}^2 \qquad L_{\text{req}} = \frac{1.138}{\pi \times 0.0424} = 8.54 \text{ m tube}$$

Chosen: 3×3 m $= 9$ m pipe.

Pressure loss calculation for double pipe heat exchanger, 3×3 m $= 9$ m tube length.
Tube side:

$$\text{Nozzle pressure loss } \Delta P_{\text{nozz}} = 1.5 \times \frac{1.248^2 \times 1100}{2} = 1285 \text{ Pa}$$

Piping pressure loss ΔP_T with friction factor $f = 0.034$ for Re $= 19,968$

$$\Delta P_\text{t} = \left(0.034 \times \frac{9}{0.0384} + 2 \right) \times \frac{1.248^2 \times 1100}{2} = 8540 \text{ Pa}$$

$$\Delta P_{\text{tot}} = 1285 + 8540 = 9825 \text{ Pa}$$

Shell side with three inlet and outlet nozzles DN 50
Nozzle flow velocity $w_{\text{St}} = 0.736$ m/s

$$\Delta P_{\text{nozz}} = 3 \times 1.5 \times \frac{0.736^2 \times 974.6}{2} = 1188 \text{ Pa}$$

$$d_\text{h}' = 0.066 - 0.0424 = 0.0236 \text{ m}$$

$$Re' = \frac{0.719 \times 0.0236}{0.385 \times 10^{-6}} = 44,073$$

Friction factor $f = 0.0278$ for $Re' = 44,073$

$$\Delta P_{\text{shell}} = 0.0278 \times \frac{9}{0.0236} \times \frac{0.719^2 \times 974.6}{2} = 2674 \text{ Pa}$$

$$\Delta P_{\text{tot}} = 2674 + 1188 = 3862 \text{ Pa}$$

Calculation of the multipipe heat exchanger:
Shell pipe 76.1×2, $d_\text{i} = 72.1$ mm with seven inner tubes 16×1
Tube side: Cross-section area $a_\text{T} = 0.0011077$ m^2

$$w_\text{t} = \frac{5.2}{0.001077 \times 3600} = 1.34 \text{ m/s} \qquad Re = \frac{1.34 \times 0.014}{2.4 \times 10^{-6}} = 7816$$

$$Pr = 21.38 \quad Nu = 87.2$$

$$\alpha_\text{i} = \frac{87.2 \times 0.4}{0.014} = 2491 \text{ W/m}^2 \text{ K} \qquad \alpha_{\text{io}} = 2491 \times \frac{14}{16} = 2179 \text{ W/m}^2 \text{ K}$$

Shell side:

$$a_{shell} = \frac{\pi}{4} \times \left(0.0721^2 - 7 \times 0.016^2\right) = 0.00267 \text{ m}^2$$

$$w_{shell} = 0.54 \text{ m/s} \qquad Pr = 2.35$$

$$d_h = \frac{0.0721^2 - 7 \times 0.016^2}{7 \times 0.016} = 0.0304 \text{ m} \qquad Re = \frac{0.54 \times 0.0304}{0.385 \times 10^{-6}} = 42,639$$

$$Nu = 0.023 \times 42,639^{0.8} \times 2.35^{0.33} = 154.2$$

$$\alpha_o = \frac{154.2 \times 0.66}{0.0304} = 3378 \text{ W/m}^2 \text{ K}$$

$$\frac{1}{U} = \frac{1}{2179} + \frac{1}{3378} + \frac{0.002}{50} + 0.0002 = 0.000995 \qquad U = 1005 \text{ W/m}^2 \text{ K}$$

$$A_{req} = \frac{51,480}{1005 \times 50.6} = 1.01 \text{ m}^2 \qquad L_{req} = \frac{1.01}{\pi \times 7 \times 0.016} = 2.9 \text{ m}$$

A heat exchanger with seven tubes of 16×1, 3 m long is required.

Pressure loss calculation for multipipe heat exchanger:
Tube side:

$$\text{Nozzles DN 32} \qquad w_{nozz} = 1.8 \text{ m/s}$$

$$\Delta P_{nozz} = 1.5 \times \frac{1.8^2 \times 1100}{2} = 2673 \text{ Pa}$$

Friction factor $f = 0.043$ for $Re' = 7816$

$$\Delta P_T = 0.043 \times \frac{2.9}{0.014} \times \frac{1.34^2 \times 1100}{2} = 8800 \text{ Pa}$$

$$\Delta P_{tot} = 2673 + 8800 = 11,473 \text{ Pa}$$

Shell side:

$$\text{Nozzles DN 50} \qquad \text{Nozzle flow velocity } w_{St} = 0.736 \text{ m/s}$$

$$\Delta P_{nozz} = \frac{1.5 \times 0.736^2 \times 974.6}{2} = 396 \text{ Pa}$$

$$d_h' = 0.0185 \text{ m} \qquad w_{shell} = 0.54 \text{ m/s}$$

$$Re' = \frac{0.54 \times 0.0185}{0.385 \times 10^{-6}} = 25,948$$

$$f = 0.032 \quad \text{for } Re' = 25,948$$

$$\Delta P_{shell} = 0.032 \times \frac{2.9}{0.0185} \times \frac{0.54^2 \times 974.6}{2} = 708 \text{ Pa}$$

$$\Delta P_{tot} = 708 + 396 = 1104 \text{ Pa}$$

11.2 HELICAL COIL HEAT EXCHANGER [3,4]

These equipments are used for low and average heat duties or low flow rates in order to achieve a higher flow velocity for the heat transfer. The heat transfer areas are greater than that of the double pipe heat exchangers, and the heat transfer is a bit better due to the centrifugal effect.

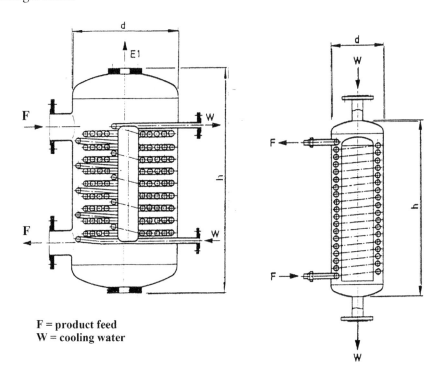

F = product feed
W = cooling water

Geometrical calculations:

Spiral total tube length L $\qquad$ $L = N \times \sqrt{(D_H \times \pi)^2 + T^2}\ (m)$

Spiral total surface area A $\qquad$ $A = \dfrac{Q}{U \times \Delta t}\ (m^2)$

Number of spiral windings N $\qquad$ $N = \dfrac{A}{\pi \times d_o \times \frac{L}{N}}$

Spiral height H $\qquad$ $H = N \times T + d_o\ (m)$

Tube-side heat transfer coefficient calculation:

$$Re > 8000$$

$$Nu = 0.023 \times Re^{0.8} \times Pr^{0.33} \qquad \alpha_i = \frac{Nu \times \lambda}{d_i} \qquad \alpha_{SP} = \alpha_i \times \left(1 + 3.5 \times \frac{d_i}{D_H}\right)$$

$$\alpha_{SPa} = \alpha_{SP} \times \frac{d_i}{d_o}$$

Calculation of the overall heat transfer coefficient:

$$\frac{1}{U} = \frac{1}{\alpha_{SPo}} + \frac{1}{\alpha_{shell}} + \frac{s_W}{\lambda_W} + f_i + f_o$$

Calculation of the shell-side flow cross-section a_{shell} and the flow velocity w_{shell}:

$$a_{shell} = \frac{\pi}{4} \times \left[\left(D_i^2 - D_K^2 \right) - \left(D_{Ho}^2 - D_{Hi}^2 \right) \right] (m^2)$$

$$w_{shell} = \frac{V_{shell}}{a_{shell} \times 3600} (m/s)$$

Calculation of α_{shell} with the hydraulic diameter D_e:

$$D_e = \frac{4 \times V_f}{\pi \times d_o \times L} = \frac{4 \times (V_a - V_c)}{\pi \times d_o \times L} (m)$$

$$V_a = \frac{\pi}{4} \times \left(D_i^2 - D_K^2 \right) \times T \times N (m^3) \qquad V_c = \frac{\pi}{4} \times d_o^2 \times L (m^3)$$

$$Re = \frac{w_{shell} \times D_e}{\nu}$$

$$Nu = 0.6 \times Re^{0.5} \times Pr^{0.31}$$

$$\alpha_{shell} = \frac{Nu \times \lambda}{D_e} (W/m^2\,K)$$

Alternative calculation of α_{shell} with the outer tube diameter d_o:

$$Re = \frac{w_{shell} \times d_o}{\nu}$$

$$Nu = 0.196 \times Re^{0.6} \times Pr^{0.33}$$

$$\alpha_{shell} = \frac{Nu \times \lambda}{d_o} (W/m^2\,K)$$

Example 4: Design of a helical coil heat exchanger

$$D_i = 0.46\,m \qquad D_K = 0.34\,m \qquad D_H = 0.4\,m$$
$$d_o = 30\,mm \qquad d_i = 25\,mm$$

$T = 45\,mm = 1.5 \times d_a$

Tube side : $1.55\,m^3/h$ $127\,°C \rightarrow 100\,°C$

Shell side : $2.3\,m^3/h$ $30\,°C \rightarrow 47\,°C$

$Q = 42,400\,W$ $LMTD = 74.9\,°C$ $F = 0.99$ $CMTD = 0.99 \times 74.9 = 74.1\,°C$

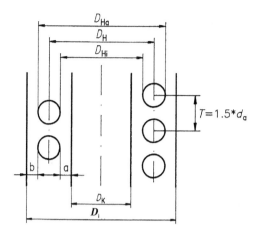

Tube-side calculation:

$$d_i = 25 \text{ mm} \qquad V = 1.55 \text{ m}^3/\text{h} \qquad \rho = 870 \text{ kg/m}^3$$
$$\nu = 0.719 \times 10^{-6} \text{ m}^2/\text{s} \qquad c = 1.16 \text{ Wh/kg K} \qquad \lambda = 0.486 \text{ W/m K}$$
$$w = 0.877 \text{ m/s}$$

$$\text{Re} = \frac{0.877 \times 0.025}{0.719 \times 10^{-6}} = 30,513 \qquad \text{Pr} = \frac{0.719 \times 10^{-6} \times 1.16 \times 870 \times 3600}{0.486} = 5.37$$

$$\text{Nu} = 0.023 \times \text{Re}^{0.8} \times \text{Pr}^{0.33} = 0.023 \times 30,513^{0.8} \times 5.37^{0.33} = 155.7$$

$$\alpha_i = \frac{155.7 \times 0.486}{0.025} = 3027 \text{ W/m}^2 \text{ K}$$

$$\alpha_{SP} = 3027 \times \left(1 + 3.5 \times \frac{0.025}{0.4}\right) = 3690 \text{ W/m}^2 \text{ K}$$

$$\alpha_{SPo} = 3690 \times \frac{25}{30} = 3075 \text{ W/m}^2 \text{ K}$$

Shell-side calculation with the hydraulic diameter D_e:

$$L = N \times \sqrt{(D_H \times \pi)^2 + T^2} = N \times \sqrt{(0.4 \times \pi)^2 + 0.045^2} = 1.257 \times N$$

$$V_a = \frac{\pi}{4} \times \left(0.46^2 - 0.34^2\right) \times 0.045 \times N = 0.0034 \times N$$

$$V_c = \frac{\pi}{4} \times 0.03^2 \times 1.257 \times N = 0.0009 \times N$$

$$V_f = V_a - V_c = (0.0034 - 0.0009) \times N = 2.504 \times 10^{-3} \times N$$

$$D_e = \frac{4 \times 2.504 \times 10^{-3} \times N}{\pi \times 0.03 \times 1.257 \times N} = 84.52 \times 10^{-3} \text{ m}$$

Calculation of *the flow cross-section* a_{shell}:

$D_{Ha} = D_i - d_a = 460 - 30 = 430$ mm

$D_{Hi} = D_K + d_a = 340 + 30 = 370$ mm

$$a_{shell} = \frac{\pi}{4} \times \left[\left(0.46^2 - 0.34^2 \right) - \left(0.43^2 - 0.37^2 \right) \right] = 0.0377 \text{ m}^2$$

Calculation of the *flow velocity* w_{shell} for $V_{shell} = 2.3$ m³/h:

$$w_{shell} = \frac{2.3}{0.0377 \times 3600} = 0.017 \text{ m/s}$$

Calculation of the *heat transfer coefficient* α_{shell} with D_e:

$$v = 1.71 \times 10^{-6} \text{ m}^2/\text{s} \qquad c = 1.16 \text{ Wh/kg K} \qquad \rho = 935 \text{ kg/m}^3 \qquad \lambda = 0.473 \text{ W/m K}$$

$$Pr = \frac{1.71 \times 10^{-6} \times 1.16 \times 935 \times 3600}{0.473} = 14.11$$

$$Re = \frac{w \times D_e}{v} = \frac{0.017 \times 0.0845}{1.71 \times 10^{-6}} = 840$$

$$Nu = 0.6 \times Re^{0.5} \times Pr^{0.31} = 0.6 \times 840^{0.5} \times 14.11^{0.31} = 39.5$$

$$\alpha_{shell} = \frac{39.5 \times 0.473}{0.0845} = 221 \text{ W/m}^2 \text{ K}$$

Calculation of the overall *heat transfer coefficient* U:

$$\alpha_{SPo} = 3075 \text{ W/m}^2 \text{ K} \qquad \alpha_{shell} = 221 \text{ W/m}^2 \text{ K}$$

$$\frac{1}{U} = \frac{1}{3075} + \frac{1}{221} + \frac{0.0025}{50} + 0.0002 = 0.0051 \qquad U = 196 \text{ W/m}^2 \text{ K}$$

$$A_{req} = \frac{42,400}{74.1 \times 196} = 2.92 \text{ m}^2 \qquad N_{req} = \frac{2.92}{\pi \times 0.03 \times 1.257} = 24.6$$

Chosen: $N = 25$

$$L = 25 \times 1.257 = 31.4 \text{ m} \qquad H = 25 \times 0.045 + 0.03 = 1.155 \text{ m}$$

Alternative calculation of α_{shell} with $d_o = 30$ mm:

$$Re = \frac{0.017 \times 0.03}{1.71 \times 10^{-6}} = 298.2$$

$$Nu = 0.196 \times Re^{0.6} \times Pr^{0.33} = 0.196 \times 298.2^{0.8} \times 14.11^{0.33} = 14.46$$

$$\alpha_{shell} = \frac{14.46 \times 0.473}{0.03} = 228 \text{ W/m}^2\text{K}$$

$$U = 201 \text{ W/m}^2 \text{ K}$$

11.3 CROSS FLOW BUNDLE [5,6,7]

Cross stream bundles are rectangular tube bundles which are predominantly used for gas cooling or heating. It will be shown how the heat transfer coefficients and pressure losses in flows through the cross stream bundles are determined.

The convective heat transfer coefficient in the tubes is treated in Chapter 3.

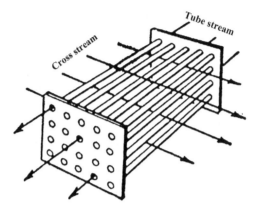

Application cases:
- air cooler for cooling or condensing of products in the tubes
- air or water preheater with hot flue gases

Due to the poor heat transfer coefficients on the gas side, finned tubes are often used to increase the efficiency.

Calculation of the heat transfer coefficients according to Grimison [5]:

$$Nu = K \times Re^M \times f_R$$

$$\alpha = \frac{Nu \times \lambda}{d} \left(W/m^2 \, K\right)$$

The constants K and M as a function of the cross and longitudinal ratio for inline and staggered arrangement are listed in Table 11.1.

As from 10 tube rows, the correction factor is $f_R = 1$.

$$\text{Cross pitch ratio} \quad s_C = \frac{P_C}{d_o}$$

$$\text{Longitudinal pitch ratio} \quad s_L = \frac{P_L}{d_o}$$

$P_C = $ Cross pitch $P_L = $ Longitudinal pitch

Table 11.1 K- and M-values for air at different cross- and longitudinal pitch ratios according to Grimison [5]

S_C	1.25		1.5		2		3	
	K	M	K	M	K	M	K	M
S_L				Inline tube banks				
1.25	0.348	0.592	0.275	0.608	0.100	0.704	0.0633	0.752
1.5	0.367	0.586	0.250	0.620	0.101	0.702	0.0678	0.744
2	0.418	0.570	0.299	0.602	0.229	0.632	0.198	0.648
3	0.290	0.601	0.357	0.584	0.374	0.581	0.286	0.608
S_L				Staggered tube banks				
0.6	—	—	—	—	—	—	0.213	0.636
0.9	—	—	—	—	0.446	0.571	0.401	0.581
1.0	—	—	0.497	0.558	—	—	—	—
1.125	—	—	—	—	0.478	0.565	0.518	0.560
1.25	0.518	0.556	0.505	0.554	0.519	0.556	0.522	0.562
1.5	0.451	0.568	0.460	0.562	0.452	0.568	0.488	0.568
2	0.404	0.572	0.416	0.568	0.482	0.556	0.449	0.570
3	0.310	0.592	0.356	0.580	0.440	0.562	0.421	0.574

For the pressure loss, the following holds:

$$\Delta P = \varsigma \times n \times \frac{w^2 \times \rho}{2} \, (\text{Pa})$$

ς = pressure loss constant n = number of tube rows
w = velocity (m/s) ρ = density (kg/m^3)

According to Jakob [7], the following holds for aligning arrangement:

$$\varsigma = \text{Re}^{-0.15} \times \left[0.176 + \frac{0.32 \times s_L}{(s_C - 1)^{(0.43 + 1.13/s_L)}} \right]$$

According to Jakob, the following holds for the staggered arrangement:

$$\varsigma = \text{Re}^{-0.16} \times \left[1 + \frac{0.47}{(s_C - 1)^{1.08}} \right]$$

Example 5: Heat transfer coefficients and the pressure drop in a cross flow bundles

air at 200 °C and 1 bar	$Pr = 0.685$	$\varrho = 0.7457$ kg/m^3
$\lambda = 0.0368$ W/mK	$\nu = 34.63$ mm^2/s	$d = 38$ mm
$n = 20$	$s_C = 1.5$	$s_L = 1.5$

The heat transfer coefficients and the pressure losses at different flow velocities w are calculated.

w (m/s)	5	10	15	20	25
Re	5486	10,973	16,460	21,946	27,433

Aligned arrangement:	$K = 0.25$	$M = 0.62$			
Nu	52	79.97	102.8	122.9	141.1
α (W/m^2 K)	52.8	81.2	104.5	124.8	143.4
ζ	0.348	0.313	0.295	0.282	0.273
ΔP (Pa)	65	233.9	495	843	1274

Staggered arrangement:	$K = 0.46$	$M = 0.562$			
Nu	58.1	85.8	107.7	126.6	143.6
α (W/m^2 K)	59	87.1	109.4	128.6	145.8
ζ	0.5027	0.449	0.421	0.4027	0.3886
ΔP (Pa)	93.7	335.5	707.5	1201.3	1811

Example 6: Calculation of the heat transfer coefficient for 18,000 m^3/h air with the physical properties of Example 5 for a tube bundle with $A_{tot} = 1$ m^2 face area with different cross and longitudinal pitches

a) $s_C = 3$ $\qquad$ $s_L = 1.5$ $\qquad$ $K = 0.0678$ $\qquad$ $M = 0.744$ $\qquad$ $A_{ges} = 1$ m^2

$$A_{free} = A_{tot} \times \frac{s_C - 1}{s_C} = 1 \times \frac{3 - 1}{3} = 0.666 \text{ m}^2$$

$$w = \frac{18,000}{0.666 \times 3600} = 7.5 \text{ m/s} \qquad Re = \frac{w \times d}{\nu} = \frac{7.5 \times 0.038}{34.63 \times 10^{-6}} = 8230$$

$Nu = 55.5$ $\qquad$ $\alpha = 56.4$ (W/m^2 K)

b) $s_C = 3$ $\qquad$ $s_L = 3$ $\qquad$ $K = 0.286$ $\qquad$ $M = 0.608$
$\quad$ $Nu = 68.7$ $\qquad$ $\alpha = 69.8$ W/m^2 K

c) $s_C = 2$ $\qquad$ $s_L = 1.5$ $\qquad$ $K = 0.101$ $\qquad$ $M = 0.702$ $\qquad$ $\alpha = 70.4$ W/m^2 K
$\quad$ $A_{frei} = 0.5$ m^2 $\qquad$ $w = 10$ m/s $\qquad$ $Re = 10,973$ $\qquad$ $Nu = 69.28$

d) $s_C = 2$ $\qquad$ $s_L = 2$ $\qquad$ $K = 0.229$ $\qquad$ $M = 0.632$
$\quad$ $Nu = 81.9$ $\qquad$ $\alpha = 83.2$ W/m^2 K

e) $s_C = 1.5$ $\qquad$ $s_L = 1.5$ $\qquad$ $K = 0.25$ $\qquad$ $M = 0.62$ $\qquad$ $\alpha = 104.5$ W/m^2 K
$\quad$ $A_{frei} = 0.333$ m^2 $\qquad$ $w = 15$ m/s $\qquad$ $Re = 16,460$ $\qquad$ $Nu = 102.8$

f) $s_C = 1.5$ $\qquad$ $s_L = 3$ $\qquad$ $K = 0.357$ $\qquad$ $M = 0.584$
$\quad$ $Nu = 103.5$ $\qquad$ $\alpha = 105$ W/m^2 K

g) $s_C = 1.25$ $\qquad$ $s_L = 1.5$ $\qquad$ $K = 0.367$ $\qquad$ $M = 0.586$ $\qquad$ $\alpha = 148.7$ W/m^2 K
$\quad$ $A_{frei} = 0.2$ m^2 $\qquad$ $w = 25$ m/s $\qquad$ $Re = 27,433$ $\qquad$ $Nu = 146.4$

h) $s_C = 1.25$ $\qquad$ $s_L = 3$ $\qquad$ $K = 0.29$ $\qquad$ $M = 0.601$
$\quad$ $Nu = 134.8$ $\qquad$ $\alpha = 137$ m^2 K

NOMENCLATURE

A_{shell} flow cross-section in the shell (m^2)

α_{io} tube-side heat transfer coefficient based on the tube outer surface area (W/m^2 K)

α_o shell-side heat transfer coefficient (W/m^2 K)

α_i tube-side heat transfer coefficient (W/m^2 K)

η dynamic viscosity (mPa s)

λ heat conductivity of the flowing product (W/m K)

λ_{wall} heat conductivity of the tube wall (W/m K)

ν kinematic viscosity (m^2/s)

ρ density (kg/m^3)

A_{req} required heat exchange area (m^2)

c specific heat capacity (Wh/kg K)

d_o tube outer diameter (m)

d_h hydraulic diameter for heat transfer coefficient calculation (m)

d'_h hydraulic diameter for pressure drop calculation (m)

D_i shell inner diameter (m)

d_i tube inner diameter (m)

f friction factor (-)

f_i, f_o fouling factors tube and shell side (m^{2}K/W)

U overall heat transfer coefficient (W/m^2 K)

L tube length (m)

LMTD logarithmic mean temperature difference (K)

n number of inner tubes (-)

Nu Nußelt number (-)

Pr Prandtl number (-)

Δp_{shell} pressure loss in the shell (Pa)

Δp_T pressure loss in the tube (Pa)

Δp_{nozz} pressure loss in the nozzle (Pa)

Q heat load (W)

Re Reynolds number (-)

V_{shell} volumetric flow in the shell (m^3/s)

V_T volumetric flow in the tube (m^3/s)

w_{shell} flow velocity in the shell (m/s)

w_T flow velocity in the tube (m/s)

w_{nozz} flow velocity in the nozzle (m/s)

z number of the 180°-turns (-)

D_i shell inner diameter

D_K core tube outer diameter

D_H average spiral diameter

d_o tube outer diameter

d_i tube inner diameter

T spiral pitch $= 1.5 \times d_o$

N number of spiral windings

D_{Ho} outer spiral diameter $= D_i - d_o$

D_{Hi} inner spiral diameter $= D_K + d_o$

a inner gap $\sim$ 15 mm or $d_o/2$

b outer gap $\sim$ 15 mm or $d_o/2$

L total tube length

REFERENCES

[1] G.P. Purohit, Thermal and hydraulic design of hairpin and finned-bundle exchangers, Chem. Eng. 90 (1983) 62–70.
[2] D.A. Donohue, Heat transfer and pressure drop in heat exchangers", Ind. Eng. Chem. 41 (1949), 2499 and Petrol.Ref. 34(1955), No. 10 and No. 11.
[3] R.K. Patil, B.W. Shende, P.K. Ghosh, Designing a helical-coil heat exchanger, Chem. Eng. 13 (12) (1982).
[4] S.S. Haraburda, Consider helical-coil heat exchangers, Chem. Eng. 102 (1995) 149–152.
[5] E.D. Grimison, Trans. ASME 59 (1937) 583–594.
[6] O.L. Pierson, Trans. ASME 59 (1937) 563.
[7] M. Jakob, Trans. ASME 60 (1938) 381–392.

CHAPTER 12

Finned Tube Heat Exchangers

Contents

12.1 WHY FINNED TUBE HEAT EXCHANGERS?

Finned tubes are used if the heat transfer coefficient on the outside of the tubes is very much lower than the heat transfer coefficient inside the tubes. The key point for the heat transfer coefficient is the heat conduction through the boundary layer δ on the tube.

$$\alpha = \frac{\lambda}{\delta} \ \left(\text{W/m}^2 \text{ K}\right)$$

In media having poor heat conductivity λ, for instance, gases or for high viscous materials with larger boundary film thickness δ, the heat load Q can be improved by an outer area enlargement.

$$Q = \alpha_o \times A_o \times \Delta t_o = \alpha_i \times A_i \times \Delta t_i$$

The required area ratio considering the fouling factors is calculated as follows:

$$\left(\frac{A_o}{A_i}\right)_{req} \geq \frac{\frac{1}{\alpha_o} + r_o}{\frac{1}{\alpha_i} + r_i}$$

© 2016 Elsevier Inc.
All rights reserved.

A_o = outer surface area of the tube (m^2/m)
A_i = inner surface area of the tube (m^2/m)
α_o = outer heat transfer coefficient (W/m^2 K)
α_i = inner heat transfer coefficient (W/m^2 K)
r_o = outer fouling factor
r_i = inner fouling factor

Example 1: Calculation of the required area ratio

$$\alpha_0 = 100 \text{ W}/\text{m}^2 \text{ K} \quad \alpha_i = 1000 \text{ W}/\text{m}^2 \text{ K} \quad r_0 = 0.0001 \quad r_i = 0.0002$$

$$\left(\frac{A_o}{A_i}\right)_{req} = \frac{\frac{1}{100} + 0.0001}{\frac{1}{1000} + 0.0002} = 8.5$$

Due to the poor outer heat transfer coefficient, the outer surface area should be larger than the inner surface tube area by 8.5!

Advantages of finned tubes
- higher heat load per m tube or per construction volume
- smaller equipment dimensions/less tubes
- smaller flow cross-section in the tubes/better heat transfer
- less pressure loss

A distinction is made between the following types of finned tubes:
- *high-finned cross-finned tubes* with finned heights of 10–16 mm
- *high-finned longitudinal-finned tubes* with finned heights of 12.7–25 mm
- *low-finned cross-finned tubes* with finned heights of 1.5–3 mm

The high-finned cross-finned tubes are used in air coolers and gas heat exchangers.

The longitudinal-finned tubes are used as vessel heater or in double pipe heat exchangers for high viscous media, for instance oil or bitumen, because—as opposed to the cross-finned tubes—the distances between the fins are much greater than the laminar boundary layer on the tube so that a flow between the fins is possible.

For a face flow length of 10 mm, the following laminar boundary layer thickness results depending on the Reynolds number Re at plates:

$$Re = 100 \quad | \quad \delta = 4.64 \text{ mm}$$
$$Re = 200 \quad | \quad \delta = 3.28 \text{ mm}$$
$$Re = 500 \quad | \quad \delta = 2.07 \text{ mm}$$

The fin spacing must be larger than the boundary layer thickness! Low-finned tubes are used to increase the efficiency in existing heat exchangers or to reduce the size of the equipment, for instance, for the refrigerant evaporation and condensation or for the decrease of the construction heights of heating bundles.

At convective heat transfer on the shell side, the Reynolds number should be >500.

12.2 WHAT PARAMETERS INFLUENCE THE EFFECTIVENESS OF FINNED TUBES?

In Figure 12.1, it can be seen how strong the outer surface area of a tube with 20 mm inner diameter increases depending on the fin height.

With increasing fin height, however, the fin efficiency η_F, which includes the temperature drop from the core tube up to the fin tip falls.

This is shown in Figure 12.2. It can also be seen how strong the heat conductivity of the fin material has an influence on the fin efficiency.

The different curves for various α_o-values in Figure 12.3 show the influence of the outer heat transfer coefficient α_o on the fin efficiency η_F.

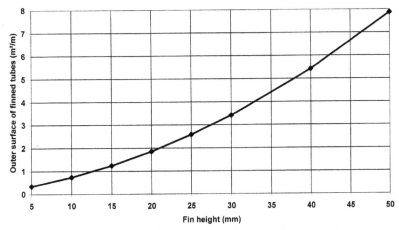

Figure 12.1 Outer tube surface area as a function of the fin height for a core tube diameter of 20 mm.

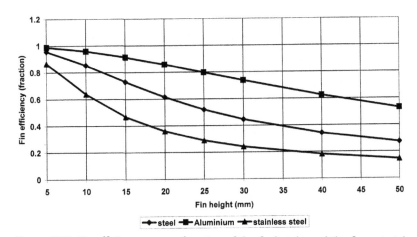

Figure 12.2 Fin efficiency η_F as a function of the fin height and the fin material.

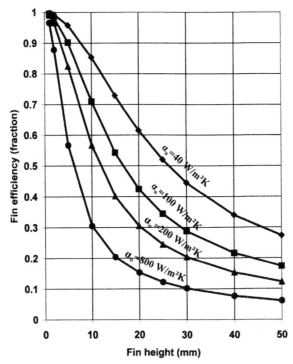

Figure 12.3 Fin efficiency η_F as a function of the fin height for different outer heat transfer coefficients of $\alpha_o = 40-800$ W/m^2 K.

With increasing fin height and a rising α_o-value, the fin efficiency falls.

Larger fin heights are only of interest with small heat transfer coefficients on the outer side of the tubes, for instance, with gases.

- Large fin heights are economic for small outer heat transfer coefficients up to $\alpha_o = 50$ W/m^2 K.
- Low-finned tubes have an advantage at higher α_o-values up to 1000 W/m^2 K.

The deterioration by the fin efficiency η_F of the effective Δt for the heat transfer is only valid for the fin surface area and not for the core tube outer surface area.

This is considered in the weighted fin efficiency η_W, which is lightly better than the fin efficiency η_F. From economic view, the weighted fin efficiency should be >80%.

The following listing shows the use of different steel-finned tubes with an example:

$$\text{Heat duty } Q = 100 \text{ kW} \qquad \Delta t = 50 \text{ K} \qquad \alpha_i = 3000 \text{ W/m}^2 \text{ K}$$

The required tube length L for the heat duty $Q = 100\,\text{kW}$ is determined for different α_o-values and finned tube types.

			$\alpha_o = 50\ \text{W/m}^2\,\text{K}$	$\alpha_o = 500\ \text{W/m}^2\,\text{K}$
Type 1:	$h_F = 16$ mm	$d_C = 38$ mm	$L = 340$ m	$L = 145$ m
Type 2:	$h_F = 10$ mm	$d_C = 20$ mm	$L = 905$ m	$L = 313$ m
Type 3:	$h_F = 1.5$ mm	$d_C = 22.2$ mm	$L = 1695$ m	$L = 317$ m
Plain tube			$L = 4348$ m	$L = 623$ m
25×2				

h_F = fin height
d_C = core tube diameter

12.3 FINNED TUBE CALCULATIONS [1–3]

12.3.1 Calculation of the fin efficiency η_F

$$\eta_F = \frac{\tanh X}{X}$$

$$\tanh X = \frac{e^X - e^{-X}}{e^X + e^{-X}}$$

$$X = h_F \times \sqrt{\frac{2 \times \alpha_o}{b_F \times \lambda_F}}$$

α_o = outer heat transfer coefficient (W/m² K)
λ_F = heat conductivity of the fin material (W/m K)
b_F = fin width (m)
h_F = fin height (m)

Correction for disk-finned tubes in which X_{DF} is inserted for the calculation of the fin efficiency η_F instead of X:

$$X_{DF} = X \times \left(1 + 0.35 \times \ln \frac{d_F}{d_C}\right)$$

d_F = fin diameter (mm)
d_C = core tube diameter (mm)

12.3.2 Calculation of the weighted fin efficiency η_W

$$\eta_F = \frac{\eta_F \times A_F + A_C}{A_o}$$

A_o = total outer surface area = $A_F + A_C$ (m²)
A_C = core tube surface area (m²)
A_F = fin surface area (m²)

Example 2: Calculation of the fin efficiency for $\alpha_o = 40$ W/m² K

$d_C = 20$ mm	$d_F = 40$ mm	$A_o = 0.55$ m²/m
$b_F = 0.3$ mm	$\lambda_F = 50$ W/m K	$d_i = 16$ mm
$h_F = 10$ mm	$A_C = 0.07$ m²/m	$A_F = 0.48$ m²/m

$$X = \sqrt{\frac{2 \times 40}{0.0003 \times 50}} \times 0.01 = 0.7303$$

$$X_{DF} = 0.7303 \times \left(1 + 0.35 \times \ln\frac{40}{20}\right) = 0.9075 \qquad \tanh X_{DF} = 0.7199$$

$$\eta_F = \frac{\tanh X}{X} = \frac{0.7199}{0.9075} = 0.7933$$

$$\eta_W = \frac{0.7933 \times 0.48 + 0.07}{0.55} = 0.82$$

Example 3: The same data as in Example 2, but $\alpha_o = 300$ W/m² K

$$X = 0.01 \times \sqrt{\frac{2 \times 300}{0.0003 \times 50}} = 2$$

$$X_{DF} = 2.48$$

$$\eta_F = 0.3968$$

$$\eta_W = \frac{0.3968 \times 0.48 + 0.07}{0.55} = 0.4736$$

Example 4: The same data as in Example 2, but $\alpha_o = 1000$ W/m² K

$$X = 0.01 \times \sqrt{\frac{2 \times 1000}{0.0003 \times 50}} = 3.65$$

$$X_{DF} = 4.5373$$

$$\eta_F = 0.22 \quad \eta_W = 0.32$$

12.3.3 Calculation of the overall heat transfer coefficient U_i for the inner tube surface area A_i without fouling

$$\frac{1}{U_i} = \frac{1}{\alpha_{oi}} + \frac{1}{\alpha_i} + \frac{s}{\lambda_F}$$

$$\alpha_{oi} = \alpha_o \times \eta_F \times \frac{A_o}{A_i} = \frac{\alpha_o}{A_i} \times (\eta_F \times A_F + A_C)$$

$$Q_i = U_i \times A_i \times \Delta t \text{ (W)}$$

α_o = outer heat transfer coefficient (W/m^2 K)
α_{oi} = outer heat transfer coefficient based on the core tube surface area
α_i = inner heat transfer coefficient (W/m^2 K)
λ_t = heat conductivity of the core tube (W/m K)
s = tube wall thickness of the core tube (m)
Δt = driving temperature gradient (K)
Q_i = heat duty in relation to the inner tube surface area (W)

Example 5: Calculation of the overall heat transfer coefficient U_i based on the tube *inner surface area A_i per m tube without fouling consideration*

α_i = 3000 W/m^2 K s = 2 mm λ_t = 50 W/m K Δt = 30 K
A_i = 0.05 m^2/m A_o = 0.55 m^2/m A_o/A_i = 11

1. α_o = 40 W/m^2 K η_F = 0.7933 η_W = 0.82

$$\alpha_{oi} = \alpha_o \times \eta_W \times \frac{A_o}{A_i} = 40 \times 0.82 \times 11 = 360 \text{ W/m}^2 \text{ K}$$

$$\alpha_{oi} = \frac{\alpha_o}{A_i} \times (\eta_F \times A_F + A_C) = \frac{40}{0.05} \times (0.7933 \times 0.48 + 0.07) = 360 \text{ W/m}^2 \text{ K}$$

$$\frac{1}{U_i} = \frac{1}{360} + \frac{1}{3000} + \frac{0.002}{50} = 0.0032$$

$$U_i = 317 \text{ W/m}^2 \text{ K}$$

$$Q_i = U_i \times A_i \times \Delta t = 317 \times 0.05 \times 30 = 475.5 \text{ W/m tube}$$

2. α_o = 300 W/m^2 K η_W = 0.4736

$$\alpha_{oi} = 300 \times 0.4736 \times 11 = 1563 \text{ W/m}^2 \text{ K}$$

$$\frac{1}{U_i} = \frac{1}{1563} + \frac{1}{3000} + \frac{0.002}{50} = 0.001 \quad U_i = 987 \text{ W/m}^2 \text{ K}$$

$$Q_i = 987 \times 30 \times 0.05 = 1480.5 \text{ W/m tube}$$

3. $\alpha_o = 1000 \text{ W/m}^2 \text{ K}$ $\eta_W = 0.32$

$$\alpha_{oi} = 1000 \times 0.32 \times 11 = 3520 \text{ W/m}^2 \text{ K}$$

$$\frac{1}{U_i} = \frac{1}{3520} + \frac{1}{3000} + \frac{0.002}{50} = 0.0007 \quad U_i = 1521 \text{ W/m}^2 \text{ K}$$

$$Q_i = 1521 \times 30 \times 0.05 = 2281.5 \text{ W/m tube}$$

12.3.4 Calculation of the overall heat transfer coefficient U_o based on the outer tube area A_o without fouling

$$\frac{1}{U_o} = \frac{1}{\alpha_{oW}} + \frac{A_o}{A_i} \times \left(\frac{s}{\lambda} + \frac{1}{\alpha_i} \right)$$

$$\alpha_{oW} = \alpha_o \times \left(1 - (1 - \eta_F) \times \frac{A_F}{A_o} \right) = \frac{\alpha_o}{A_o} \times (\eta_F \times A_F + A_C) = \alpha_o \times \eta_W$$

$$Q_o = U_o \times A_o \times \Delta t \ (\text{W})$$

α_o = outer heat transfer coefficient (W/m^2 K)

α_{oW} = weighted outer heat transfer coefficient under consideration of the fin efficiency for the fin surface area

Q_a = heat duty based on the outer tube surface area (W)

$$U_o = U_i \times \frac{A_i}{A_o} \qquad U_i = U_o \times \frac{A_o}{A_i}$$

Example 6: Calculation of the overall heat transfer coefficient U_a based on the outer tube surface area per m tube without fouling consideration

Data as in Example 5

1. $\alpha_o = 40 \text{ W/m}^2 \text{ K}$ $\eta_F = 0.7933$

$$\alpha_{oW} = 40 \times \left(1 - \left(1 - 0.7933 \times \frac{0.48}{0.55} \right) \right) = 32.8 \text{ W/m}^2 \text{ K}$$

$$\frac{1}{U_o} = \frac{1}{32.8} + \frac{0.55}{0.05} \times \left(\frac{0.002}{50} + \frac{1}{3000} \right) = 0.0346$$

$$U_a = 28.82 \text{ W/m}^2 \text{ K} \qquad F_o = 0.55 \text{ m}^2/\text{m tube}$$

$$Q_o = 28.82 \times 30 \times 0.55 = 475.5 \text{ W/m tube}$$

Conversion:

$$U_i = U_o \times A_o/A_i = 28.82 \times 11 = 317 \text{ W/m}^2 \text{ K}$$

$$Q_i = 317 \times 30 \times 0.05 = 475.5 \text{ W/m tube}$$

2. $\alpha_A = 300 \text{ W/m}^2 \text{ K}$ $\eta_R = 0.3963$

$$\alpha_{oW} = 300 \times \left(1 - (1 - 0.3963) \times \frac{0.48}{0.55}\right) = 141.9 \text{ W/m}^2 \text{ K}$$

$$\frac{1}{U_o} = \frac{1}{141.9} + \frac{0.55}{0.05} \times \left(\frac{0.002}{50} + \frac{1}{3000}\right) = 0.0112$$

$$U_o = 89.72 \text{ W/m}^2 \text{ K}$$
$$Q_a = 89.72 \times 30 \times 0.55 = 1480.4 \text{ W/m tube}$$

Conversion : $U_i = 89.72 \times 11 = 986.9 \text{ W/m}^2 \text{ K}$
$$Q_i = 986.9 \times 30 \times 0.05 = 1480.35 \text{ W/m tube}$$

12.3.5 Calculation of the overall heat transfer coefficient U_o based on the finned outer surface area A_o considering the fouling factors r_o and r_i

$$\frac{1}{U_o} = \frac{\dfrac{1}{\alpha_o} + r_o}{\eta_F} + \frac{A_o}{A_i} \times \left(\frac{1}{\alpha_i} + r_i \times \frac{s_W}{\lambda_W}\right)$$

$$\frac{1}{U_o} = \frac{1}{\alpha_{oW}} + \frac{r_o}{\eta_W} + \frac{A_o}{A_i} \times \left(\frac{1}{\alpha_i} + \frac{s_W}{\lambda_W} + r_i\right)$$

$$Q_a = U_a \times A_A \times \Delta t$$

Example 7: Calculation of the overall heat transfer coefficient U_a considering fouling

$\alpha_o = 800 \text{ W/m}^2 \text{ K}$ $\alpha_i = 6000 \text{ W/m}^2 \text{ K}$

$A_o = 0.207 \text{ m}^2/\text{m}$ $A_o/A_i = 3.27$ $\eta_W = 0.9368$

$r_o = r_i = 0.00015$ $s_W = 1 \text{ mm}$ $\lambda_W = 50 \text{ W/m K}$

$\Delta t = 25 \text{ K}$

$$\alpha_{oW} = \alpha_o \times \eta_W = 800 \times 0.9368 = 749.4 \text{ W/m}^2 \text{ K}$$

$$\frac{1}{U_o} = \frac{1}{749.4} + \frac{0.00015}{0.9368} + 3.27 \times \left(\frac{1}{6000} + \frac{0.001}{50} + 0.00015\right) = 0.0026$$

$$U_o = 385.3 \text{ W/m}^2 \text{ K}$$

$$Q_o = U_o \times A_o \times \Delta t = 385.3 \times 0.207 \times 25 = 1994 \text{ W/m tube}$$

12.3.6 Calculation of the overall heat transfer coefficient U_i for the inner tube surface area A_i considering the fouling factors r_o and r_i

$$\frac{1}{U_i} = \frac{1}{\alpha_{oi}} + \left(\frac{r_o}{\eta_W} \times \frac{A_i}{A_o}\right) + \frac{1}{\alpha_i} + r_i + \frac{s_W}{\lambda_W}$$

$$Q_i = U_i \times A_i \times \Delta t$$

Example 8: Calculation of U_i considering fouling data as in Example 7

$$\alpha_{oi} = \alpha_o \times \eta_W \times \frac{A_o}{A_i} = 800 \times 0.9368 \times 3.27 = 2450.6 \text{ W/m}^2 \text{ K}$$

$$\frac{1}{U_i} = \frac{1}{2450.6} + \left(\frac{0.00015}{0.9368} \times \frac{1}{3.27}\right) + \frac{1}{6000} + 0.00015 + \frac{0.001}{50} = 0.0008$$

$$U_i = 1259.9 \text{ W/m}^2 \text{ K}$$

$$Q_i = U_i \times A_i \times \Delta t = 1259.9 \times 0.0633 \times 25 = 1994 \text{ W/m tube}$$

Conversion:

$$U_o = U_i \times \frac{A_i}{A_o} = 1259.9 \times \frac{1}{3.27} = 385.3 \text{ W/m}^2 \text{ K}$$

$$Q_o = U_o \times A_o \times \Delta t = 385.3 \times 0.207 \times 25 = 1994 \text{ W/m tube}$$

12.3.7 Fouling and Temperature Gradient

A further advantage of fin tubes is the smaller temperature drop by foulings on the enlarged outer tube surface area F_A. The fouling does not have such a strong effect as with plain tubes.

$$\Delta t_{ro} = r_o \times \frac{q}{A_o} \ (K) \quad \Delta t_{ri} = r_i \times \frac{q}{A_i} \ (K)$$

Δt_{ro} = temperature drop by fouling on A_o
Δt_{ri} = temperature drop by fouling on A_i
q = heat flux density (W/m tube)

Since the surface area of finned tubes is much larger than the surface area of the inner tube the fouling effect on the outer side is considerably less.

Example 9: Calculation of the temperature drop by fouling

$A_o = 0.207 \text{ m}^2/\text{m}$	$A_i = 0.0638 \text{ m}^2/\text{m}$
$r_o = r_i = 0.0002$	$U_o = 531 \text{ W/m}^2 \text{ K}$ $\Delta t = 25 \text{ K}$

$$Q = U_o \times A_o \times \Delta t = 531 \times 0.207 \times 25 = 2748 \text{ W/m tube}$$

$$\Delta t_{ro} = 0.0002 \times \frac{2748}{0.207} = 2.6 \text{ K}$$

$$\Delta t_{ri} = 0.0002 \times \frac{2748}{0.0638} = 8.6 \text{ K}$$

The temperature drop on the large outer surface area is much smaller!

The temperature gradient at the outer side is also reduced due to the larger outer surface area.

The temperature difference between the wall and the medium is smaller so that a more careful heating with less Δt is possible, or at high temperatures on the tube outer side the material is treated carefully.

Example 10: Calculation of the temperature gradients for plain tube and finned tube

$Q = 500$ kW	$\alpha_o = 800$ W/m^2 K	$\alpha_i = 6000$ W/m^2 K	
$r_o = r_i = 0.00015$	$s_W = 1$ mm	$\lambda_W = 50$ W/m K	$\Delta t = 25$ K

1. Plain tube: 25 × 1 $F_o = 0.07854$ m^2/m $A_o/A_i = 1.087$

$$\frac{1}{U_o} = \frac{1}{800} + 0.00015 + \frac{0.001}{50} + 1.087 \times \left(\frac{1}{6000} + 0.00015 \right)$$

$$U_o = 566.8 \text{ W/m}^2 \text{ K}$$

$$q = U_o \times \Delta t = 566.8 \times 25 = 14{,}170 \text{ W/m}^2$$

$$\Delta t_{\alpha o} = \frac{14170}{800} = 17.71 \text{ °K}$$

$$\Delta t_{ro} = 0.00015 \times 14170 = 2.13 \text{ °K}$$

$$\Delta t_{\alpha i} = \frac{14170}{6000} \times 1.087 = 2.57 \text{ °K}$$

$$\Delta t_{ri} = 0.00015 \times 14170 \times 1.087 = 2.31 \text{ °K}$$

$$\Delta t_W = \frac{0.001}{50} \times 14170 = 0.28 \text{ °K}$$

2. Finned tube Trufin S/T

$A_o = 0.207$ m^2/m $A_o/A_i = 3.27$ $\eta_W = 0.9368$

$$\frac{1}{U_o} = \frac{\frac{1}{800} + 0.00015}{0.9368} + 3.27 \times \left(\frac{1}{6000} + 0.00015 + \frac{0.001}{50} \right)$$

$$U = 385.3 \text{ W/m}^2 \text{ K}$$

$$q = U_o \times \Delta t = 385.3 \times 25 = 9632.5 \text{ W/m}^2$$

$$\Delta t_{\alpha o} = \frac{9632.5}{800 \times 0.9368} = 12.85\,^{\circ}K$$

$$\Delta t_{ro} = \frac{0.00015 \times 0632.5}{0.9368} = 1.55\,^{\circ}K$$

$$\Delta t_{\alpha i} = \frac{9632.5}{6000} \times 3.27 = 5.25\,^{\circ}K$$

$$\Delta t_{ri} = 0.00015 \times 9632.5 \times 3.27 = 4.72\,^{\circ}K$$

$$\Delta t_W = \frac{0.001}{50} \times 9632.5 \times 3.27 = 0.63\,^{\circ}K$$

$\Delta t_{\alpha o}$ = temperature gradient for α_o
$\Delta t_{\alpha i}$ = temperature gradient for α_i
Δt_{ro} = temperature gradient for the outer fouling r_o
Δt_{ri} = temperature gradient for the inner fouling r_i
Δt_W = temperature gradient for the heat conductivity through the wall

12.3.8 Comparison of the specific heat duties $U_i \times A_i$ (W/m K) of different tubes

In Figure 12.4, the specific heat duties per m tube of different tubes as a function of the outer heat transfer coefficient are shown.

The curves are valid for an inner heat transfer coefficient of $\alpha_i = 3000$ W/m^2 K in the tube.

- Plain steel tubes 25×2 made, $A_o/A_i = 1.19$
- Low-finned Trufin tubes 22.2×1.65 with 1.5 mm steel fin height, $A_o/A_i = 3.6$
- High-finned Applifin tubes 20×2 with 10 mm fin height, $A_o/A_i = 11$

  It can be clearly seen that by the finning, the heat duty can be increased.

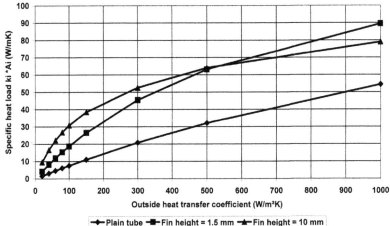

Figure 12.4 Heat load (W/m K) as a function of the outer heat transfer coefficients.

The advantage is seen with the high-finned tubes in the area of lower outer heat transfer coefficients and also the possibilities for an increase of the efficiency by the use of low-finned tubes in the area of higher outer heat transfer coefficients.

12.4 APPLICATION EXAMPLES

From the discussions so far, it follows that finned tubes are preferably applied if a low heat transfer coefficient should be compensated by a large surface area. From the following calculated examples, the preferred application of different fin tube types for specific tasks results:

- From Example 11, the advantages of low-finned tubes for heating bundles for the evaporation if low construction heights and low bundle widths are desired result.
- Example 12 emphasizes the advantages of longitudinal fin tubes for heating coils for heating storage tanks, because the heat transfer by natural convection is very poor.
- From Example 13, it is clear that the high-finned cross-finned tubes are suitable for the gas cooling or heating.
- In Example 14, it is shown that double-tube heat exchangers with finned tubes and multitube heat exchangers are very suitable for high viscous and gaseous media, because in small room large duties are possible [4].

Example 11: Evaporator heating bundle for a distillation still without consideration of fouling and the heat conduction resistance of the tube wall

$Q = 500$ kW $\qquad \alpha_i = 6000$ W/m^2 K $\quad \alpha_o = 800$ W/m^2 K $\quad \Delta t = 25$ K

1. With plain tubes 25 × 1 $\quad L = 4$ m $\qquad A_o = 0.0785$ m^2/m

$$\frac{1}{U_o} = \frac{1}{6000} \times \frac{25}{23} + \frac{1}{800} \quad U_o = 699 \text{ W/m}^2 \text{ K}$$

$$A_{req} = \frac{Q}{U \times \Delta t} = \frac{500{,}000}{699 \times 25} = 28.6 \text{ m}^2 = 364 \text{ m tube DN 25}$$

Arrangement: 91 Rohre, 4 m long, pitch 32 mm
Bundle width $B = 90 \times 32 + 25 = 2905$ mm

2. With low-finned Trufin tubes S/T 60-197042

$d_o = 25.4$ mm $\qquad h_F = 1.5$ mm $\qquad b_F = 0.3$ mm $\qquad \lambda_F = 50$ W/m K
$A_o = 0.207$ m^2/m $\qquad A_o/A_i = 3.27$ $\qquad d_F/d_C = 1.135$
$X = 0.4898$ $\qquad X_{DF} = 0.5116$ $\qquad \eta_F = 0.921$ $\qquad \eta_W = 0.9368$

$$\alpha_{oW} = \eta_W \times \alpha_o = 0.9368 \times 800 = 749.4 \text{ W/m}^2\text{K}$$

$$\frac{1}{U_o} = \frac{1}{749.4} + \frac{1}{6000} \times 3.27 \quad U_o = 532 \text{ W/m}^2\text{K}$$

$$A_{req} = \frac{500{,}000}{532 \times 25} = 37.6 \text{ m}^2 = 182 \text{ m tube DN 25}$$

Arrangement: 46 tubes, 4 m long, pitch 32 mm
Bundle width $B = 45 \times 32 + 25.4 = 1465.4$ mm

3. Longitudinal-finned tubes, $d_C = 25.4$ mm, with 20 longitudinal fins 12.7 mm high,

$b_F = 0.81$ mm $A_o = 0.5869$ m^2/m $A_o/A_i = 8.838$

$X = 2.52$ $\eta_F = 0.391$ $\eta_W = 0.4559$ $\alpha_{oW} = 364.7$ W/m^2 K

$$\frac{1}{U_o} = \frac{1}{364.7} + \frac{1}{6000} \times 8.838 \quad U_o = 237 \text{ W/m}^2 \text{ K}$$

$$A_{req} = \frac{500,000}{237 \times 25} = 84.4 \text{ m}^2 = 144 \text{ m tube}$$

Arrangement: 36 tubes, 4 m long, pitch 66 mm
Bundle width $B = 35 \times 66 + 50.8 = 2360.8$ mm
Corollary: The smallest bundle results by applying the low-finned tubes!

Example 12: Heating bundle for the heating of a product in a storage tank with 10 m diameter without considering the fouling and the heat conduction resistance of the tube wall

$Q = 200$ kW $\alpha_i = 6000$ W/m^2 K $\alpha_o = 50$ W/m^2 K $\Delta t = 50$ K

Tube data as in Example 11.

1. With plain tubes 25 × 1 $A_o = 0.0785$ m^2

$$\frac{1}{U_o} = \frac{1}{6000} \times \frac{25}{23} + \frac{1}{50} \quad U_o = 49.5 \text{ W/m}^2 \text{ K}$$

$$A_{req} = \frac{200,000}{49.5 \times 50} = 80.8 \text{ m}^2 = 1029 \text{ m pipe DN 25}$$

Arrangement: 128 tubes, 8 m long, single pass

2. With Trufin-finned tubes $A_o = 0.207$ m^2/m $A_o/A_i = 3.27$

$X = 0.122$ $X_{DF} = 0.1279$ $\eta_F = 0.9945$ $\eta_W = 0.9957$
$\alpha_{AW} = 0.9957 \times 50 = 49.8$ W/m^2 K

$$\frac{1}{U_o} = \frac{1}{49.8} + \frac{1}{6000} \times 3.27 \quad U_o = 48.5 \text{ W/m}^2 \text{ K}$$

$$A_{req} = \frac{200,000}{48.5 \times 50} = 82.5 \text{ m}^2 = 398 \text{ m pipe}$$

Arrangement: 50 tubes, 8 m long, single pass

3. With longitudinal-finned tubes

$d_C = 25.4$ mm $h_F = 12.7$ mm $A_o = 0.5869$ m^2/m $A_o/A_i = 8.838$
$\eta_F = 0.885$ $\eta_W = 0.897$ $\alpha_{oW} = 0.897 \times 50 = 44.85$ W/m^2 K

$$\frac{1}{U_o} = \frac{1}{44.85} + 8.838 \times \frac{1}{6000} \quad U_o = 42 \text{ W/m}^2 \text{ K}$$

$$A_{req} = \frac{200,000}{42 \times 50} = 95.2 \text{ m}^2 = 162 \text{ m pipe}$$

Arrangement: 20 tubes, 8 m long, single pass
Recommendation: Heating bundle consisting of longitudinal-finned tubes

Example 13: Gas cooling in a cross flow bundle with cooling water in the tubes

$\alpha_i = 5000$ W/m^2 K in the tubes

Gas flow rate $V_{Shell} = 904{,}000$ m^3/h $Q = 8$ Mio W $\Delta t = 40$ K

Allowable pressure loss $= 3.8$ mbar

Allowable bundle width $= 6$ m

Properties of the gases: $\varrho = 0.885$ kg/m^3 $\lambda = 0.0332$ W/m K
                                  $\nu = 25$ mm^2/s   Pr $= 0.68$

1. With plain tubes 38×3.6 without fouling

Estimation of the required area A with $U = 90$ W/m^2 K:

$$A = \frac{Q}{U \times \Delta t} = \frac{8 \times 10^6}{90 \times 40} = 2222 \text{ m}^2$$

$$L_{req} = 19{,}000 \text{ m} = 3166 \text{ tubes with } L = 6 \text{ m}$$

Due to the low allowable pressure losses, an aligning arrangement with $P_C = P_L = 2 \times da = 76$ mm is chosen.

$P_C =$ cross pitch of the tubes

$P_L =$ longitudinal pitch of the tubes

Arrangement: 100 tubes one over the other, 6 m long, 36 tube rows one behind the other, aligned arrangement

Total tube length $= 100 \times 36 \times 6 = 21{,}600$ m   Bundle length $L = 6$ m

Total surface area $= 2578$ m^2 Bundle height $H = 7.6$ m

Bundle cross-sectional area $A_{bundle} = 6 \times 7.6 = 45.6$ m^2 (width $\times$ height)

Free flow cross-sectional area A_{free} for the gas through the tube bundle:

$$A_{free} = \frac{P_C - d_o}{P_C} \times H \times L = \frac{76 - 38}{76} \times 7.6 \times 6 = 22.8 \text{ m}^2$$

$$w_{gas} = \frac{904{,}000}{22.8 \times 3600} = 11 \text{ m/s}$$

$$\text{Re} = \frac{w_{gas} \times d_o}{\nu} = \frac{11 \times 0.038}{25 \times 10^{-6}} = 16{,}720$$

Pressure loss calculation according to Jakob (Section 11.3): $\zeta = 0.189$

$$\Delta P = \frac{w^2 \times \rho}{2} \times n_R \times \zeta = \frac{11 \times 0.88}{2} \times 36 \times 0.189 = 362 \text{ Pa}$$

Heat transfer calculation according to Grimison (Section 11.3):

$$\text{Nu} = 0.229 \times 16{,}720^{0.632} \times 1 = 106.8$$

$$\alpha = \frac{106.8 \times 0.0332}{0.038} = 93.3 \text{ W/m}^2 \text{ K}$$

Calculation of the overall heat transfer coefficient for steam heating with $\alpha_i = 5000$ W/m^2 K:

$$\alpha_{io} = 5000 \times \frac{30.8}{38} = 4428 \text{ W/m}^2 \text{ K}$$

$$\frac{1}{U} = \frac{1}{4428} + \frac{1}{93.3} + \frac{0.0036}{50} = 0.0011$$

$U = 90.7 \ \text{W/m}^2 \ \text{K}$

Heat duty Q: $Q = k \times A \times \Delta t = 90.7 \times 2578 \times 40 = 9.35 \ \text{W}$

Required heat duty $Q_{req} = 8$ Mio W   Fouling reserve: 16.9%
Bundle dimensions: Height $= 7.6$ m   Width $= 6$ m   Depth $= 2.7$ m

2. With finned tubes Applifin 95,725 without fouling

Outer surface area $A_o = 1.5 \ \text{m}^2/\text{m}$ $d_C = 25.4$ mm $d_F = 57.2$ mm
Cross pitch $P_C = 57.2$ mm   Longitudinal pitch $P_L = 49.5$ mm, staggered

Arrangement: 153 tubes one over the other, 6 m long, five rows one behind the other

Total tube length $L_{tot} = 153 \times 6 \times 5 = 4590$ m tube   Total surface area $A = 6885 \ \text{m}^2$
$A_{bundle} = 7.6 \times 6 = 45.6 \ \text{m}^2$ $f_{proj} = 0.03 \ \text{m}^2/\text{m}$ tube $A_{free} = 45.6 \times 6 \times 0.03 = 18 \ \text{m}^2$

$$w = \frac{904,000}{3600 \times 18} = 13.95 \ \text{m/s} \quad Re = \frac{13.95 \times 0.0254}{25 \times 10^{-6}} = 14,174$$

Pressure loss calculation with $\zeta = 0.8$ from manufacturer data

$$\Delta P = \frac{13.95^2 \times 0.88}{2} \times 5 \times 0.8 = 343 \ \text{Pa}$$

Calculation of the heat transfer coefficients according to manufacturer data:

$$\frac{Nu}{Pr^{1/3}} = 0.37 \times Re^{0.553} \times f_R = 0.37 \times 14,174^{0.553} \times 1 = 73.1$$

$$Nu = 73.1 \times 0.68^{0.33} = 64.4$$

$$\alpha_o = \frac{64.4 \times 0.0332}{0.0254} = 84.2 \ \text{W/m}^2 \ \text{K}$$

Calculation of the fin efficiency:

$$X = 0.0159 \times \sqrt{\frac{2 \times 84.2}{50 \times 0.0004}} = 1.459 \quad X_{DF} = 1.8735 \quad \eta_F = 0.5092$$

$$\eta_W = \frac{0.5092 \times 1.46 + 0.04}{1.5} = 0.5222$$

$$\alpha_{oW} = 0.5222 \times 84.2 = 44 \ \text{W/m}^2 \ \text{K}$$

Calculation of the overall heat transfer coefficient:

$\alpha_i = 4000 \ \text{W/m}^2 \ \text{K}$ $\alpha_{oW} = 44 \ \text{W/m}^2 \ \text{K} =$ effective α-value shell side

$$\frac{1}{U} = \frac{1}{44} + 23.5 \times \left(\frac{0.0025}{50} + \frac{1}{4000} \right) = 0.0298 \quad U = 33.6 \ \text{W/m}^2\text{K}$$

Heat duty $Q = 33.6 \times 6885 \times 40 = 9.25$ Mio W

Required heat load: $Q_{req} = 8$ Mio W   Reserve: 15.7%
Bundle dimensions: Height: 7.6 m   Width: 6 m   Depth: 255.2 mm

Comparison between plain tube and finned tube:

	Plain tube	Finned tube
Height (m)	7.6	7.6
Width (m)	6,-	6
Depth (m)	2.7	0.255
Tube length (m)	21,600	4590
Surface area (m^2)	2578	6885

Corollary: By the use of finned tubes, the tube length can be reduced from 21,600 to 4590 m and the bundle depth from 2.7 to 0.26 m.

Example 14: Double pipe oil cooler with an inner longitudinal-finned tube

Pipe shell with $D_i = 77.9$ mm $A_o/A_i = 5.5$

Inner tube with longitudinal fins: $d_o = 48.2$ mm, fin height 12.7 mm, surface area 0.76 m^2/m

Tube-side product: cooling water

Required heat load $Q_{req} = 7$ kW	Effective $\Delta t = 23$ °C
$s_W = 2$ mm $\qquad \lambda_W = 59$ W/m K	$r_a = r_i = 0.0001$

Shell side product rate $V_{Shell} = 3.5$ m^3/h Flow cross-section $A_{free} = 0.0026$ m^2

Flow velocity $w_{Shell} = 0.374$ m/s.

Oil data: $\qquad \rho = 846$ kg/m^3 $\qquad c = 0.58$ Wh/kg K $\quad \lambda = 0.131$ W/m K $\quad \nu = 19$ mm^2/s

$$\text{Pr} = \frac{3600 \times 19 \times 10^{-6} \times 0.58 \times 846}{0.131} = 256$$

$$\text{Re} = \frac{0.374 \times 0.0104}{19 \times 10^{-6}} = 204.7 \rightarrow \text{Laminar Flow}$$

Calculation of the heat transfer coefficient:

$$\text{Nu} = 1.86 \times \left(204.7 \times 256 \times \frac{0.0104}{6}\right)^{1/3} = 8.36$$

$$\alpha_o = \frac{8.36 \times 0.131}{0.0104} = 105 \text{ W/m}^2 \text{ K}$$

$$X = 0.0127 \times \sqrt{\frac{2 \times 105}{0.001 \times 50}} = 0.823 \quad \eta_F = 0.822$$

$$\eta_W = \frac{0.822 \times 0.61 + 0.15}{0.76} = 0.857$$

$$\alpha_{oW} = 0.857 \times 105 = 90 \quad \text{W/m}^2 \text{ K}$$

Tube side: 3 m^3/h cooling water.

$d_i = 44.6$ mm $\qquad w_T = 0.54$ m/s $\qquad \text{Pr} = 6.94 \qquad \lambda = 0.604$ W/m K

$$\text{Re} = \frac{0.54 \times 0.0442}{1 \times 10^{-6}} = 23,868$$

$$\text{Nu} = 0.023 \times 23,868^{0.8} \times 6.94^{1/3} = 139.45$$

$$\alpha_T = \frac{139.45 \times 0.604}{0.0442} = 1905 \text{ W/m}^2 \text{ K}$$

Calculation of the overall heat transfer coefficient U_o:

$$\frac{1}{U_o} = \frac{1}{\alpha_{oW}} + \frac{r_o}{\eta_W} + \frac{A_o}{A_i} \times \left(\frac{1}{\alpha_i} + \frac{s_W}{\lambda_W} + r_i\right)$$

$$\frac{1}{U_o} = \frac{1}{90} + \frac{0.0001}{0.857} + 5.5 \times \left(\frac{0.002}{59} + \frac{1}{1905} + 0.0001\right) \quad U_o = 68.1 \ \text{W/m}^2 \ \text{K}$$

$$\text{Required area } A_{req} = \frac{7000}{23 \times 68.1} = 4.47 \ \text{m}^2$$

$$\text{Required tube length } L_{req} = \frac{4.47}{0.76} = 5.88 \ \text{m pipe}$$

Alternative calculation for a multipipe heat exchanger with plain tubes:
Shell diameter $D_i = 84$ mm with seven inner tubes 20×2
Shell side: 3.5 m^3/h oil

Flow cross-section	De = 0.0304 m	Surface area = 0.4398 m^2/m
$f_{Shell} = 0.0033 \ \text{m}^2$		
$w_{Shell} = 0.295$ m/s	Re = 471	

$$Nu = 1.86 \times \left(471 \times 256 \times \frac{0.0304}{6}\right) = 15.78$$

$$\alpha_o = \frac{15.78 \times 0.131}{0.0304} = 68 \ \text{W/m}^2\text{K}$$

Tube side: 3 m^3/h cooling water $w_T = 0.59$ m/s   Re = 18,009
Nu = 111.3 $\alpha_i = 4202 \ \text{W/m}^2 \ \text{K}$ $\alpha_{io} = 3362 \ \text{W/m}^2 \ \text{K}$

Calculation of the overall heat transfer coefficient U:

$$\frac{1}{U} = \frac{1}{3362} + \frac{1}{68} + \frac{0.002}{50} + 0.0002$$

$$U = 65.6 \ \text{W/m}^2 \ \text{K}$$

$$\text{Required area } A_{req} = \frac{7000}{65.6 \times 23} = 4.64 \ \text{m}^2$$

$$\text{Required length } L_{req} = \frac{4.64}{0.4398} = 10.6 \ \text{m}$$

REFERENCES

[1] D.Q. Kern, A.D. Kraus, Extended Surface Heat Transfer, McGraw-Hill, N.Y., 1972.
[2] Th.E. Schmidt, Kältetechnik 18 (4) (1966) 135−138.
[3] W.M. Kays, A.L. London, Compact Heat Exchangers, McGraw-Hill, N.Y., 1964.
[4] G.P. Purohit, Thermal and hydraulic design of hairpin and finned-bundle exchangers, Chem. Eng., 90 (1983) 62−70.

INDEX

'Note: Page numbers followed by "f" indicate figures, "t" indicate tables, and "b" indicate boxes.'

Printed in the United States
By Bookmasters